Relly Victoria Petrescu
&
Florian Ion Petrescu

LOCKHEED
MARTIN
COLOR

2012

Scientific reviewer:
Dr. Veturia CHIROIU

Honorific member of
Technical Sciences Academy of Romania (ASTR)
PhD supervisor in Mechanical Engineering

Copyright

Title book: Lockheed Martin Color

Author book: Relly Victoria Petrescu, Florian Ion Petrescu

© 2012, Florian Ion PETRESCU

petrescuflorian@yahoo.com

ISBN 978-1-4818-2781-2

Welcome! A Short Book Description:

Lockheed Martin (NYSE: LMT) is an American global aerospace, defense, security, and advanced technology company with worldwide interests. It was formed by the merger of Lockheed Corporation with Martin Marietta in March 1995. It is headquartered in Bethesda, Maryland, in the Washington Metropolitan Area. Lockheed Martin employs 123,000 people worldwide. Robert J. Stevens is the current Chairman and Chief Executive Officer.

Lockheed Martin is one of the world's largest defense contractors; In 2009, 74% of Lockheed Martin's revenues came from military sales. It received 7.1% of the funds paid out by the Pentagon.

Lockheed Martin operates in four business segments. These comprise, with respective percentages of 2009 total net sales of $45.2 billion, Aeronautics (27%), Electronic Systems (27%), Information Systems & Global Solutions (27%), and Space Systems (19%). In 2009 US Government contracts accounted for $38.4 billion (85%), foreign government contracts $5.8 billion (13%), and commercial and other contracts for $900 million (2%). In both 2009 and 2008 the company topped the list of US Federal Contractors.

The company has received the Collier Trophy six times. Most recently (in 2001) for being part of developing the X-35/F-35B LiftFan Propulsion System, and again in 2006 for leading the team that developed the F-22 Raptor fighter jet. Lockheed Martin is currently developing the F-35 Lightning II.

Merger talks between Lockheed Corporation and Martin Marietta began in March 1994, with the companies announcing their $10 billion planned merger on August 30, 1994. The deal was finalized on March 15, 1995 when the two companies' shareholders approved the merger. The segments of the two companies not retained by the new company formed the basis for the present L-3 Communications, a mid-size defense contractor in its own right. Lockheed Martin later spun off the materials company Martin Marietta Materials.

Both companies contributed important products to the new portfolio. Lockheed products included the Trident missile, P-3 Orion, F-16 Fighting Falcon, F-22 Raptor, C-130 Hercules, A-4AR Fightinghawk and the DSCS-3 satellite. Martin Marietta products included Titan rockets, Sandia National Laboratories (management contract acquired in 1993), Space Shuttle External Tank, Viking 1 and Viking 2 landers, the Transfer Orbit Stage (under subcontract to Orbital Sciences Corporation) and various satellite models.

On April 22, 1996, Lockheed Martin completed the acquisition of Loral Corporation's defense electronics and system integration businesses for $9.1 billion, the deal having been announced in January. The remainder of Loral became Loral Space & Communications.

Lockheed Martin abandoned plans for a $8.3 billion merger with Northrop Grumman on July 16, 1998, due to government concerns over the potential strength of the new group; Lockheed/Northrop would have had control of 25% of the Department of Defense's procurement budget.

Lockheed Martin provided NASA with measurements in US Customary force units when metric was expected, resulting in the loss of the Mars Climate Orbiter at a cost of $125 million. The cost for spacecraft development was $193.1 million.

In May 2001, Lockheed Martin sold Lockheed Martin Control Systems to BAE Systems. On November 27, 2000, Lockheed completed the sale of its Aerospace Electronic Systems business to BAE Systems for $1.67 billion, a deal announced in July 2000. This group encompassed Sanders Associates, Fairchild Systems, and Lockheed Martin Space Electronics & Communications.

In 2001, Lockheed Martin won the contract to build the F-35 Lightning II; this was the largest fighter aircraft procurement project since the F-16, with an initial order of 3,000 worth some $200 billion before export orders.

In 2001, Lockheed Martin settled a nine year investigation conducted by NASA's Office of Inspector General with the assistance of the Defense Contract Audit Agency. The company paid the United States government $7.1 million based on allegations that its predecessor, Lockheed Engineering Science Corporation, submitted false lease costs claims to NASA.

On May 12, 2006, The Washington Post reported that when Robert Stevens took control of Lockheed Martin in 2004, he faced the dilemma that within 10 years 100,000 of the about 130,000 Lockheed Martin employees – more than three-quarters – would be retiring.

On August 31, 2006, Lockheed Martin won a $3.9 billion contract from NASA to design and build the CEV capsule, also known as Orion – the next spacecraft for human flight – for the Ares I rocket in the Constellation Program.

On August 13, 2008, Lockheed Martin acquired the government business unit of Nantero, Inc., a company that had developed methods and processes for incorporating carbon nanotubes in next-generation electronic devices. In 2009, Lockheed Martin bought Unitech.

On November 18, 2010, Lockheed Martin announced that it would be closing its Eagan, MN location by 2013 in order to reduce costs and optimize capacity at their locations nationwide.

In January 2011, Lockheed Martin agreed to pay the US Government $2 million to settle allegations that the company submitted false claims on a U.S. government contract for that amount. The allegations came from a contract with the Naval Oceanographic Office Major Shared Resource Center in Mississippi.

On May 28, 2011 it was reported that a cyber-attack using previously stolen EMC files had broken through to sensitive materials at the contractor. It is unclear if the Lockheed incident is the specific prompt whereby on June 1, 2011, the new United States military strategy, makes explicit that a cyberattack is casus belli for a traditional act of war.

On July 10, 2012, Lockheed Martin announced it was cutting its workforce by 740 workers as a measure of cost reduction to remain competitive necessary for future growth.

On August 2, 2012, the Vice President for Business Development, George Standridge stated that his company has offered 6 more C-130J aircraft to the Indian Air Force, for which discussions are underway with the Indian Government.

On November 27, 2012, Lockheed Martin announced that Marillyn Hewson will become the corporation's chief executive officer on January 1, 2013.

CONTENT

Lockheed Martin Aeronautics

Lockheed Martin Aeronautics Company is a major unit of Lockheed Martin with headquarters at Air Force Plant 4 in Fort Worth, Texas.

Lockheed Martin Aeronautics is also based in Marietta, Georgia and Palmdale, California. Palmdale is home to the Advanced Development Programs (ADP), informally known as the "Skunk Works". Various subassemblies are produced at locations in Florida, Mississippi, Pennsylvania and West Virginia.

The company draws upon the history of the former Lockheed and Martin Marietta corporations. While the formation of Lockheed Martin in 1995 was a merger of equals, by far the greatest contribution to Lockheed Martin Aeronautics was the product portfolio of Lockheed. This included the C-5, C-130 and C-141 transports as well as the F-2, F-16 (purchased from General Dynamics), F-117, F-22 and F-35 Lightning II.

The most important project by far to Lockheed Martin Aeronautics is the F-35 Lightning II (JSF). Worth a potential $200bn the initial order book is approximately 3,000 excluding almost guaranteed export orders. The F-22 air dominance fighter, while smaller in terms of aircraft ordered, is still of great importance to Lockheed Martin (and partner Boeing).

C-130 Hercules

First developed by the Lockheed Corporation in the 1950s, the Hercules is still produced today as the C-130J Super Hercules. The aircraft is produced at Lockheed Martin's Marietta, Georgia facility.

The Lockheed C-130 Hercules is a four-engine turboprop military transport aircraft designed and built originally by Lockheed, now Lockheed Martin. Capable of using unprepared runways for takeoffs and landings, the C-130 was originally designed as a troop, medical evacuation, and cargo transport aircraft. The versatile airframe has found uses in a variety of other roles, including as a gunship (AC-130), for airborne assault, search and rescue, scientific research support, weather reconnaissance, aerial refueling, maritime patrol and aerial firefighting. It is now the main tactical airlifter for many military forces worldwide. Over 40 models and variants of the Hercules serve with more than 60 nations.

The C-130 entered service with U.S. in the 1950s, followed by Australia and others. During its years of service, the Hercules family has participated in countless military, civilian and humanitarian aid operations. The family has the longest continuous production run of any military aircraft in history. In 2007, the C-130 became the fifth aircraft—after the English Electric Canberra, Boeing B-52 Stratofortress, Tupolev Tu-95, and Boeing KC-135 Stratotanker—to mark 50 years of continuous use with its original primary customer, in this case, the United States Air Force. The C-130 is also the only military aircraft to remain in continuous production for 50 years with its original customer, as the updated C-130J Super Hercules.

C-130 Hercules

The Korean War, which began in June 1950, showed that World War II-era piston-engine transports—Fairchild C-119 Flying Boxcars, Douglas C-47 Skytrains and Curtiss C-46 Commandos—were inadequate for modern warfare. Thus on 2 February 1951, the United States Air Force issued a General Operating Requirement (GOR) for a new transport to Boeing, Douglas, Fairchild, Lockheed, Martin, Chase Aircraft, North American, Northrop, and Airlifts Inc. The new transport would have a capacity for 92 passengers, 72 combat troops or 64 paratroopers in a cargo compartment that was approximately 41 feet (12 m) long, 9 feet (2.7 m) high, and 10 feet (3.0 m) wide. Unlike transports derived from passenger airliners, it was to be designed from the ground-up as a combat transport with loading from a hinged loading ramp at the rear of the fuselage. This innovation for military cargo aircraft was first pioneered, known as the Trapoklappe, on the WW II German Junkers Ju 252 and Ju 352 "Hercules" transport prototypes.

The Hercules resembled a larger four-engine brother to the C-123 Provider with a similar wing and cargo ramp layout that evolved from the Chase XCG-20 Avitruc, which in turn, was first designed and flown as a cargo glider in 1947. The Boeing C-97 also had a rear ramp, which made it possible to drive vehicles onto the plane (also possible with forward ramp on a C-124). But the ramp on the Hercules was also used to airdrop cargo, which included low-altitude extraction for Sheridan tanks and even dropping large improvised "daisy cutter" bombs.

A key feature was the introduction of the T56 turboprop, first developed specifically for the C-130. At the time, the turboprop was a new application of turbine engines that used exhaust gases to turn a shafted propeller, which offered greater range at propeller-driven speeds compared to pure turbojets, which were faster but thirstier. As was the case on helicopters of that era, such as the UH-1 Huey, turboshafts produced much more power for their weight than piston engines. Lockheed would subsequently use the same engines and technology in the Lockheed L-188 Electra. That aircraft failed financially in its civilian configuration but was successfully adapted into the Lockheed P-3 Orion maritime patrol and submarine attack aircraft where the efficiency and endurance at high propeller speeds of turboprops excelled.

The new Lockheed cargo plane design possessed a range of 1,100 nmi (1,300 mi; 2,000 km), takeoff capability from short and unprepared strips, and the ability to fly with one engine shut down. Fairchild, North American, Martin and Northrop declined to participate. The remaining five companies tendered a total of 10 designs: Lockheed two, Boeing one, Chase three, Douglas three, and Airlifts Inc. one. The contest was a close affair between the lighter of the two Lockheed (preliminary project designation L-206) proposals and a four-turboprop Douglas design.

The Lockheed design team was led by Willis Hawkins, starting with a 130 page proposal for the Lockheed L-206. Hall Hibbard, Lockheed vice president and chief engineer, saw the proposal and directed it to Kelly Johnson, who did not care for the low-speed, unarmed aircraft, and remarked, "If you sign that letter, you will destroy the Lockheed Company." Both Hibbard and Johnson signed the proposal and the company won the contract for the now-designated Model 82 on 2 July 1951.

The first flight of the YC-130 prototype was made on 23 August 1954 from the Lockheed plant in Burbank, California. The aircraft, serial number 53-3397, was the second prototype but the first of the two to fly. The YC-130 was piloted by Stanley Beltz and Roy Wimmer on its 61-minute flight to Edwards Air Force Base; Jack Real and Dick Stanton served as flight engineers. Kelly Johnson flew chase in a P2V Neptune.

After the two prototypes were completed, production began in Marietta, Georgia, where over 2,300 C-130s have been built through 2009.

The initial production model, the C-130A, was powered by Allison T56-A-9 turboprops with three-blade propellers. Deliveries began in December 1956, continuing until the introduction of the C-130B model in 1959. Some A models were re-designated C-130D after being equipped with skis. The newer C-130B had ailerons with increased boost—3,000 psi (21 MPa) versus 2,050 psi (14 MPa)—as well as uprated engines and four-bladed propellers that were standard until the J-model's introduction.

As the C-130A became operational with Tactical Air Command (TAC), the C-130's lack of range became apparent and additional fuel capacity was added in the form of external pylon-mounted tanks at the end of the wings.

The C-130B model was developed to complement the A models that had previously been delivered, and incorporated new features, particularly increased fuel capacity in the form of auxiliary tanks built into the center wing section and an AC electrical system. Four-bladed Hamilton Standard propellers replaced the Aero Product three-bladed propellers that distinguished the earlier A-models. An electronic reconnaissance variant of the C-130B was designated C-130B-II. A total of 13 aircraft were converted. The C-130B-II was distinguished by its false external wing fuel tanks, which were disguised signals intelligence (SIGINT) receiver antennas. These pods were slightly larger than the standard wing tanks found on other C-130Bs. Most aircraft featured a swept blade antenna on the upper fuselage, as well as extra wire antennas between the vertical fin and upper fuselage not found on other C-130s. Radio call numbers on the tail of these aircraft were regularly changed so as to confuse observers and disguise their true mission.

The extended range C-130E model entered service in 1962 after it was developed as an interim long-range transport for the Military Air Transport Service. Essentially a B-model, the new designation was the result of the installation of 1,360 US gal (5,150 L) Sargent Fletcher external fuel tanks under each wing's midsection and more powerful Allison T56-A-7A turboprops. The hydraulic boost pressure to the ailerons was reduced back to 2050 psi as a consequence of the external tanks' weight in the middle of the wingspan. The E model also featured structural improvements, avionics upgrades and a higher gross weight. Australia took delivery of 12 C130E Hercules during 1966–67 to supplement the 12 C-130A models already in service with the RAAF. Sweden and Spain fly the TP-84T version of the C-130E fitted for aerial refueling capability.

A Michigan Air National Guard C-130E dispatches its flares during a low-level training mission

The KC-130 tankers, originally C-130Fs procured for the US Marine Corps (USMC) in 1958 (under the designation GV-1) are equipped with a removable 3,600 US gal (13,626 l) stainless steel fuel tank carried inside the cargo compartment. The two wing-mounted hose and drogue aerial refueling pods each transfer up to 300 US gal per minute (19 l per second) to two aircraft simultaneously, allowing for rapid cycle times of multiple-receiver aircraft formations, (a typical tanker formation of four aircraft in less than 30 minutes). The US Navy's C-130G has increased structural strength allowing higher gross weight operation.

The C-130H model has updated Allison T56-A-15 turboprops, a redesigned outer wing, updated avionics and other minor improvements. Later H models had a new, fatigue-life-improved, center wing that was retro-fitted to many earlier H-models. The H model remains in widespread use with the U.S. Air Force (USAF) and many foreign air forces. Initial deliveries began in 1964 (to the RNZAF), remaining in production until 1996. An improved C-130H was introduced in 1974, with Australia purchasing 12 of type in 1978 to replace the original 12 C-130A models, which had first entered RAAF Service in 1958.

The United States Coast Guard employs the HC-130H for long range search and rescue, drug interdiction, illegal migrant patrols, homeland security, and logistics.

Royal Australian Air Force C-130H, 2007

C-130H models produced from 1992 to 1996 were designated as C-130H3 by the USAF.

The 3 denoting the third variation in design for the H series. Improvements included ring laser gyros for the INUs, GPS receivers, a partial glass cockpit (ADI and HSI instruments), a more capable APN-241 color radar, night vision device compatible instrument lighting, and an integrated radar and missile warning system. The electrical system upgrade included Generator Control Units (GCU) and Bus Switching units (BSU) to provide stable power to the more sensitive upgraded components.

The equivalent model for export to the UK is the C-130K, known by the Royal Air Force (RAF) as the Hercules C.1. The C-130H-30 (Hercules C.3 in RAF service) is a stretched version of the original Hercules, achieved by inserting a 100 in (2.54 m) plug aft of the cockpit and an 80 in (2.03 m) plug at the rear of the fuselage. A single C-130K was purchased by the Met Office for use by its Meteorological Research Flight, where it was classified as the Hercules W.2. This aircraft was heavily modified (with its most prominent feature being the long red and white striped atmospheric probe on the nose and the move of the weather radar into a pod above the forward fuselage). This aircraft, named Snoopy, was withdrawn in 2001 and was then modified by Marshall of Cambridge Aerospace as flight-test bed for the A400M turbine engine, the TP400. The C-130K is used by the RAF Falcons for parachute drops. Three C-130K (Hercules C Mk.1P) were upgraded and sold to the Austrian Air Force in 2002.

Royal Air Force C-130K (C.3)

The MC-130E Combat Talon was developed for the USAF during the Vietnam War to support special operations missions throughout Southeast Asia, and spawned a family of special missions aircraft. 37 of the earliest models currently operating with the United States Special Operations Command are scheduled to be replaced by new-production MC-130J versions. The EC-130 and EC-130H Compass Call versions are also Special variants but are assigned to Air Combat Command (ACC). The AC-130 gunship was first developed during the Vietnam War to provide close air support and other ground-attack duties.

The HC-130 is a family of long-range search and rescue variants used by the USAF and the U.S. Coast Guard. Equipped for deep deployment of Pararescuemen (PJs), survival equipment, and aerial refueling of combat rescue helicopters, HC-130s are usually the on-scene command aircraft for combat SAR missions. Early versions were equipped with the Fulton surface-to-air recovery system, designed to pull a person off the ground using a wire strung from a helium balloon. The John Wayne movie The Green Berets features its use. The Fulton system was later removed when aerial refueling of helicopters proved safer and more versatile. The movie The Perfect Storm depicts a real life SAR mission involving aerial refueling of a New York Air National Guard HH-60G by a New York Air National Guard HC-130P.

USAF HC-130P refuels a HH-60G Pave Hawk helicopter

The C-130R and C-130T are U.S. Navy and USMC models, both equipped with underwing external fuel tanks. The USN C-130T is similar, but has additional avionics improvements. In both models, aircraft are equipped with Allison T56-A-16 engines. The USMC versions are designated KC-130R or KC-130T when equipped with underwing refueling pods and pylons and are fully night vision system compatible.

The RC-130 is a reconnaissance version. A single example is used by the Islamic Republic of Iran Air Force, the aircraft having originally been sold to the former Imperial Iranian Air Force.

The Lockheed L-100 (L-382) is a civilian variant, equivalent to a C-130E model without military equipment. The L-100 also has two stretched versions.

In the 1970s, Lockheed proposed a C-130 variant with turbofan engines rather than turboprops, but the U.S. Air Force preferred the takeoff performance of the existing aircraft. In the 1980s, the C-130 was intended to be replaced by the Advanced Medium STOL Transport project. The project was canceled and the C-130 has remained in production.

In the 1990s, the improved C-130J Super Hercules was developed by Lockheed (later Lockheed Martin). This model is the newest version and the only model in production. Externally similar to the classic Hercules in general appearance, the J model has new turboprop engines, six-bladed propellers, digital avionics, and other new systems.

In 2000, Boeing was awarded a US$1.4 billion contract to develop an Avionics Modernization Program kit for the C-130.

The program was beset with delays and cost overruns until project restructuring in 2007. On 2 September 2009, Bloomberg news reported that the planned Avionics Modernization Program (AMP) upgrade to the older C-130s would be dropped to provide more funds for the F-35, CV-22 and airborne tanker replacement programs.

However, in June 2010, Department of Defense approved funding for the initial production of the AMP upgrade kits. Under the terms of this agreement, the USAF has cleared Boeing to begin low-rate initial production (LRIP) for the C-130 AMP. A total of 198 aircraft are expected to feature the AMP upgrade. The current cost per aircraft is US$14 million although Boeing expects that this price will drop to US$7 million for the 69th aircraft.

Operational history

The first production aircraft, C-130As were first delivered beginning in 1956 to the 463d Troop Carrier Wing at Ardmore AFB, Oklahoma and the 314th Troop Carrier Wing at Sewart AFB, Tennessee. Six additional squadrons were assigned to the 322d Air Division in Europe and the 315th Air Division in the Far East. Additional aircraft were modified for electronics intelligence work and assigned to Rhein-Main Air Base, Germany while modified RC-130As were assigned to the Military Air Transport Service (MATS) photo-mapping division.

In 1958, a U.S. reconnaissance C-130A-II was shot down over Armenia by MiG-17s.

Australia became the first non-American force to operate the C-130A Hercules with 12 examples being delivered from late 1958. These aircraft were fitted with AeroProducts three-blade, 15-foot diameter propellers. The Royal Canadian Air Force became another early user with the delivery of its first B-model in 1960.

A Hercules deploying flares

In 1963, a Hercules achieved and still holds the record for the largest and heaviest aircraft to land on an aircraft carrier. During October and November that year, a USMC KC-130F (BuNo 149798), loaned to the U.S. Naval Air Test Center, made 29 touch-and-go landings, 21 unarrested full-stop landings and 21 unassisted take-offs on the USS Forrestal at a number of different weights. The pilot, LT (later RADM) James H. Flatley III, USN, was awarded the Distinguished Flying Cross for his role in this test series. The tests were highly successful, but the idea was considered too risky for routine "Carrier Onboard Delivery" (COD) operations. Instead, the C-2 Greyhound was developed as a dedicated COD aircraft. The Hercules used in the test, most recently in service with Marine Aerial Refueler Squadron 352 (VMGR-352) until 2005, is now part of the collection of the National Museum of Naval Aviation at NAS Pensacola, Florida.

In 1964, C-130 crews from the 6315th Operations Group at Naha Airbase, Okinawa commenced forward air control (FAC; "Flare") missions over the Ho Chi Minh Trail in Laos supporting USAF strike aircraft. In April 1965 the mission was expanded to North Vietnam where C-130 crews led formations of B-57 bombers on night reconnaissance/strike missions against communist supply routes leading to South Vietnam. In early 1966 Project Blind Bat/Lamplighter was established at Ubon RTAFB, Thailand. After the move to Ubon the mission became a four-engine FAC mission with the C-130 crew searching for targets then calling in strike aircraft. Another little-known C-130 mission flown by Naha-based crews was Operation Commando Scarf, which involved the delivery of chemicals onto sections of the Ho Chi Minh Trail in Laos that were designed to produce mud and landslides in hopes of making the truck routes impassable.

In November 1964, on the other side of the globe, C-130Es from the 464th Troop Carrier Wing but loaned to 322d Air Division in France, flew one of the most dramatic missions in history in the former Belgian Congo. After a Congolese rebel group named "Simba" took white residents of the city of Stanleyville hostage, the U.S. and Belgium developed a joint rescue mission that used the C-130s to airlift and then drop and air-land a force of Belgian paratroopers to rescue the hostages. Two missions were flown, one over Stanleyville and another over Paulis during Thanksgiving weeks. The headline-making mission resulted in the first award of the prestigious MacKay Trophy to C-130 crews.

In the Indo-Pakistani War of 1965, the Pakistan Air Force modified/improvised several aircraft for use as heavy bombers, and attacks were made on Indian bridges and troop concentrations with some successes. No aircraft were lost in the operations, though one was slightly damaged.

In October 1968 a C-130Bs from the 463rd Tactical Airlift Wing dropped a pair of M-121 10,000 pound bombs that had been developed for the massive B-36 bomber but had never been used. The U.S. Army and U.S. Air Force resurrected the huge weapons as a means of clearing landing zones for helicopters and in early 1969 the 463rd commenced Commando Vault missions. Although the stated purpose of COMMANDO VAULT was to clear LZs, they were also used on enemy base camps and other targets.

During the late 1960s, the U.S. was eager to get information on Chinese nuclear capabilities. After the failure of the Black Cat Squadron to plant operating sensor pods near the Lop Nur Nuclear Weapons Test Base using a Lockheed U-2, the CIA developed a plan, named Heavy Tea, to deploy two battery-powered sensor pallets near the base. To deploy the pallets, a Black Bat Squadron crew was trained in the U.S. to fly the C-130 Hercules. The crew of 12, led by Col Sun Pei Zhen, took off from Takhli Royal Thai Air Force Base in an unmarked U.S. Air Force C-130E on 17 May 1969. Flying for six and a half hours at low altitude in the dark, they arrived over the target and the sensor pallets were dropped by parachute near Anxi in Gansu province. After another six and a half hours of low altitude flight, they arrived back at Takhli. The sensors worked and uploaded data to a U.S. intelligence satellite for six months, before their batteries wore out. The Chinese conducted two nuclear tests, on 22 September 1969 and 29 September 1969, during the operating life of the sensor pallets. Another mission to the area was planned as Operation Golden Whip, but was called off in 1970. It is most likely that the aircraft used on this mission was either C-130E serial number 64-0506 or 64-0507 (cn 382-3990 and 382-3991). These two aircraft were delivered to Air America in 1964. After being returned to the U.S. Air Force sometime between 1966 and 1970, they were assigned the serial numbers of C-130s that had been destroyed in accidents. 64-0506 is now flying as 62-1843, a C-130E that crashed in Vietnam on 20 December 1965 and 64-0507 is now flying as 63-7785, a C-130E that had crashed in Vietnam on 17 June 1966.

The A-model continued in service through the Vietnam War, where the aircraft assigned to the four squadrons at Naha AB, Okinawa and one at Tachikawa Air Base, Japan performed yeoman's service, including operating highly classified special operations missions such as the BLIND BAT FAC/Flare mission and FACT SHEET leaflet mission over Laos and North Vietnam. The A-model was also provided to the South Vietnamese Air Force as part of the Vietnamization program at the end of the war, and equipped three squadrons based at Tan Son Nhut AFB. The last operator in the world is the Honduran Air Force, which is still flying one of five A model Hercules (FAH 558, c/n 3042) as of October 2009. As the Vietnam War wound down, the 463rd Troop Carrier/Tactical Airlift Wing B-models and A-models of the 374th Tactical Airlift Wing were transferred back to the United States where most were assigned to Air Force Reserve and Air National Guard units.

Another prominent role for the B model was with the United States Marine Corps, where Hercules initially designated as GV-1s replaced C-119s. After Air Force C-130Ds proved the type's usefulness in Antarctica, the U.S. Navy purchased a number of B-models equipped with skis that were designated as LC-130s. C-130B-II electronic reconnaissance aircraft were operated under the SUN VALLEY program name primarily from Yokota Air Base, Japan. All reverted to standard C-130B cargo aircraft after their replacement in the reconnaissance role by other aircraft.

The C-130 was also used in the 1976 Entebbe raid in which Israeli commando forces carried a surprise assault to rescue 103 passengers of an airliner hijacked by Palestinian and German terrorists at Entebbe Airport, Uganda. The rescue force — 200 soldiers, jeeps, and a black Mercedes-Benz (intended to resemble Ugandan Dictator Idi Amin's vehicle of state) — was flown over 2,200 nmi (2,532 mi; 4,074 km) almost entirely at an altitude of less than 100 ft (30 m) from Israel to Entebbe by four Israeli Air Force (IAF) Hercules aircraft without mid-air refueling (on the way back, the planes refueled in Nairobi, Kenya).

During the Falklands War (Spanish: Guerra de las Malvinas) of 1982, Argentine Air Force C-130s undertook highly dangerous, daily re-supply night flights as blockade runners to the Argentine garrison on the Falkland Islands. They also performed daylight maritime survey flights. One was lost during the war. Argentina also operated two KC-130 tankers during the war, and these refueled both the A-4 Skyhawks and Navy Dassault-Breguet Super Étendards; some C-130s were modified to operate as bombers with bomb-racks under their wings. The British also used RAF C-130s to support their logistical operations.

During the Gulf War of 1991 (Operation Desert Storm), the C-130 Hercules was used operationally by the U.S. Air Force, U.S. Navy and U.S. Marine Corps, along with the air forces of Australia, New Zealand, Saudi Arabia, South Korea and the UK. The MC-130 Combat Talon variant also made the first attacks using the largest conventional bombs in the world, the BLU-82 "Daisy Cutter" and GBU-43/B Massive Ordnance Air Blast bomb, also known as the MOAB. Daisy Cutters were used to clear landing zones and to eliminate mine fields. The weight and size of the weapons make it impossible or impractical to load them on conventional bombers. The GBU-43/B MOAB is a successor to the BLU-82 and can perform the same function, as well as perform strike functions against hardened targets in a low air threat environment.

Since 1992, two successive C-130 aircraft named Fat Albert have served as the support aircraft for the U.S. Navy Blue Angels flight demonstration team. Fat Albert I was a TC-130G (15189), while Fat Albert II is a C-130T (164763). Although Fat Albert supports a Navy squadron, it is operated by the U.S. Marine Corps (USMC) and its crew consists solely of USMC personnel. At some air shows featuring the team, Fat Albert takes part, performing flyovers. Until 2009, it also demonstrated its jet-assisted takeoff (JATO) capabilities; these ended due to dwindling supplies of rockets.

The AC-130 also holds the record for the longest sustained flight by a C-130. From 22 to 24 October 1997, two AC-130U gunships flew 36.0 hours nonstop from Hurlburt Field Florida to Taegu (Daegu), South Korea while being refueled seven times by KC-135 tanker aircraft. This record flight shattered the previous record longest flight by over 10 hours while the 2 gunships took on 410,000 lb (190,000 kg) of fuel. The gunship has been used in every major U.S. combat operation since Vietnam, except for Operation El Dorado Canyon, the 1986 attack on Libya.

During the invasion of Afghanistan in 2001 and the ongoing support of the International Security Assistance Force (Operation Enduring Freedom), the C-130 Hercules has been used operationally by Australia, Belgium, Canada, Denmark, France, Italy, the Netherlands, New Zealand, Norway, Portugal, South Korea, Spain, the UK and the United States.

During the 2003 invasion of Iraq (Operation Iraqi Freedom), the C-130 Hercules was used operationally by Australia, the UK and the United States. After the initial invasion, C-130 operators as part of the Multinational force in Iraq used their C-130s to support their forces in Iraq.

USMC C-130T *Fat Albert* performing a RATO

Since 2004, the Pakistan Air Force has employed C-130s in the War in North-West Pakistan. Some variants had forward looking infrared (FLIR Systems Star Safire III EO/IR) sensor balls, to enable close tracking of Islamist militants.

An RAF C-130 was shot down on 30 January 2005, when an Iraqi insurgent brought it down firing with a ZU-23 anti-aircraft artillery gun while the plane was flying at 164 ft (50 m) after it had dropped SAS special forces paratroopers.

On 20 November 2011, the Australian government announced plans to donate four ex-RAAF C-130 Hercules aircraft to Indonesia for humanitarian and disaster relief work. The Hercules aircraft require an estimated A$25 million in maintenance to restore them to airworthiness.

The U.S. Forest Service developed the Modular Airborne FireFighting System for the C-130 in the 1970s, which allows regular aircraft to be temporarily converted to an airtanker for fighting wildfires. In the late 1980s, 22 retired USAF C-130As were removed from storage at Davis-Monthan Air Force Base and transferred to the U.S. Forest Service who then sold them to six private companies to be converted into air tankers (see U.S. Forest Service airtanker scandal). After one of these aircraft crashed due to wing separation in flight as a result of fatigue stress cracking, the entire fleet of C-130A air tankers was permanently grounded in 2004 (see 2002 airtanker crashes). C-130s have been used to spread chemical dispersants onto the massive oil slick in the Gulf Coast in 2010.

A C-130E fitted with a MAFFS dropping fire retardant

C-130s from the: U.S., Canada, Australia and Israel (foreground to background)

C-130H Hercules flight deck

Cargo compartment of a Swedish Air Force C-130

Philippine Air Force and Army servicemen unload a C-130 of supplies for transfer to waiting U.S. helicopters for delivery to Panay Island.

Lockheed Martin C-130J Super Hercules

The Lockheed Martin C-130J "Super" Hercules is a four-engine turboprop military transport aircraft. The C-130J is a comprehensive update of the venerable Lockheed C-130 Hercules, with new engines, flight deck, and other systems.

The Hercules family has the longest continuous production run of any military aircraft in history. During more than 50 years of service, the family has participated in military, civilian, and humanitarian aid operations.

The Hercules has outlived several planned successor designs, most notably the Advanced Medium STOL Transport contestants.

Fifteen nations have placed orders for a total of 300 C-130Js, of which 250 aircraft have been delivered as of February 2012.

An HC-130J from U.S. Coast Guard Air Station Elizabeth City flies along the coast of Elizabeth City, North Carolina.

The C-130J is the newest version of the Hercules and the only model still in production. Externally similar to the classic Hercules in general appearance, the J-model features considerably updated technology. These differences include new Rolls-Royce AE 2100 D3 turboprops with Dowty R391 composite scimitar propellers, digital avionics (including head-up displays (HUDs) for each pilot), and reduced crew requirements. These changes have improved performance over its C-130E/H predecessors, such as 40% greater range, 21% higher maximum speed, and 41% shorter takeoff distance. The J-model is available in a standard-length or stretched -30 variant.

The C-130J's crew includes two pilots and one loadmaster (no navigator or flight engineer). The United States Marine Corps utilizes a crew chief for expeditionary operations. Its cargo compartment is approximately 41 feet (12.5 m) long, 9 feet (2.74 m) high, and 10 feet (3.05 m) wide, and loading is from the rear of the fuselage. The aircraft can also be configured with the "enhanced cargo handling system". The system consists of a computerized loadmaster's station from which the user can remotely control the under-floor winch and also configure the flip-floor system to palletized roller or flat-floor cargo handling. Initially developed for the USAF, this system enables rapid role changes to be carried out and so extends the C-130J's time available to complete taskings.

Lockheed Martin received the launch order for the J-model from the RAF, which ordered 25 aircraft, with first deliveries beginning in 1999 as Hercules C4 (C-130J-30) and Hercules C5 (C-130J). The standard C-130J had a flyaway cost of US$62 million in 2008.

In mid-June 2008, the United States Air Force awarded a $470 million contract to Lockheed Martin for six modified KC-130J aircraft for use by the Air Force and Special Operations Command. The contract led to C-130J variants that will replace aging HC-130s and MC-130s. The HC-130J Combat King II personnel recovery aircraft completed developmental testing on 14 March 2011. The final test point was air-to-air refueling, and was the first ever boom refueling of a C-130 where the aircraft's refueling receiver was installed during aircraft production. This test procedure also applied to the MC-130J Combat Shadow II aircraft in production for Air Force Special Operations Command.

With the addition of the Marine Corps's ISR / Weapon Mission Kit, the KC-130J tanker variant will be able to serve as an overwatch aircraft and can deliver ground support fire in the form of Hellfire or Griffin missiles, precision-guided bombs, and eventually 30mm cannon fire in a later upgrade. This capability, designated as "Harvest HAWK" (Hercules Airborne Weapons Kit), can be used in scenarios where precision is not a requisite, such as area denial. The aircraft retains its original capabilities in refueling and transportation. The entire system can be removed within a day if necessary.

A KC-130J showing the AN/AAQ-30 Targeting Sight and AGM-114 Hellfire on the left wing in Afghanistan, 2011.

The Super Hercules has been used extensively by the USAF and USMC in Iraq. Canada has also deployed their CC-130J aircraft to Afghanistan. The Royal Canadian Air Force announced 22 August 2011, that the Tactical Airlift Unit will keep flying ISAF missions until at least mid-November.

C-130Js from several countries have been deployed in support of the US Operation Odyssey Dawn and NATO's Operation Unified Protector during the 2011 Libyan civil war.

The Modular Airborne FireFighting System (MAFFS) is a self-contained unit used for aerial firefighting that can be loaded onto a C-130 Hercules, which then allows the aircraft to be used as an air tanker against wildfires. This allows the U.S. Forest Service (USFS) to utilize military aircraft from the Air National Guard and Air Force Reserve to serve as an emergency backup resource to the civilian air tanker fleet. The latest generation MAFFS II system was used for the first time on a fire in July 2010, using the C-130J Super Hercules. The 146th Airlift Wing was the first to transition to the MAFFS II system in 2008, and it remains the only unit flying the new system on the C-130J aircraft.

The largest operator of the new model is the U.S. Air Force, which has ordered the aircraft in increasing numbers. Current operators of the C-130J are the USAF (to include the Air Force Reserve Command and the Air National Guard), United States Marine Corps (being their fourth variant after KC-130F, KC-130R and KC-130T,) United States Coast Guard, Royal Air Force, Royal Canadian Air Force, Royal Australian Air Force, Danish Air Force, Royal Norwegian Air Force, Indian Air Force and the Italian Air Force. As of July 2010, a total of 200 units have been produced of the 284 on order at that time.

A C-130J Super Hercules is cleaned in the wash system at Keesler Air Force Base, Mississippi.

Australia was the second international customer for the C-130J-30, with an initial order of twelve aircraft. An additional order for two more aircraft was planned, but cancelled with the purchase of a fifth Boeing C-17 Globemaster III.

Lockheed Martin offered to lease four C-130Js to the German Luftwaffe in 2005. Germany was awaiting the replacement of its Transall C-160 by Airbus A400Ms, but that was planned for 2010. The deal would have filled in the gap in airlift capability, but the offer was declined.

The Royal Norwegian Air Force ordered four C-130J-30s in 2007 to replace six aging C-130Hs in need of additional repairs. Aircraft were delivered from November 2008 to 2010. One of these was lost in March 2012.

The Canadian Forces signed a US$1.4 billion contract with Lockheed Martin for seventeen new C-130J-30s on 16 January 2008, as part of the procurement process to replace the existing C-130E and H models. The C-130J will be officially designated CC-130J Hercules in Canadian service. The first C-130J was delivered to CFB Trenton on 4 June 2010. The final C-130J was delivered on 11 May 2012.

The Indian Air Force purchased six C-130J-30s in early 2008 at a cost of up to US$1.059 billion for its special operations forces in a package deal with the US government under its Foreign Military Sales (FMS) program. India has options to buy six more aircraft. The Indian government decided not to sign the Communications Interoperability and Security Memorandum of Agreement (CISMOA), which resulted in the exclusion of high precision GPS and other sensitive equipment. However the IAF added similar equipment produced indigenously , to the aircraft after delivery. In October 2011, India announced its intent to exercise the option for the six additional aircraft, following the favorable performance of the C-130J in the 2011 Sikkim earthquake relief operations. In July 2012 the U.S. accepted India's request for sale of six more C-130Js through the FMS program.

The Iraqi Air Force ordered six C-130J-30s in July 2008.

Qatar ordered four C-130Js in October 2008, along with spare parts and training for the Qatar Emiri Air Force. The contract is worth a total of US$393.6 million and deliveries are scheduled to begin in 2011.

In June 2009, Lockheed Martin said that both the UK and France had asked for technical details on the C-130J as an alternative to the troubled Airbus A400M.

The United Arab Emirates Air Force announced an order for twelve C-130J transports at the 2009 IDEX, with an announced value of US$1.3 billion. The United Arab Emirates requested 12 C-130Js through a Direct Commercial Sale in December 2009, with logistics support, training and related systems to be provided through a Foreign Military Sales program. A contract with Lockheed Martin has not been signed.

The Israeli Air Force is seeking to purchase nine C-130J-30s. In April 2010 Israel ordered one C-130J-30 with delivery in 2013, and was in contract talks for two more aircraft in June 2010. An option for a second C-130J-30 was exercised on 8 April 2011, along with planning and advance long lead procurement of aircraft components to support the third C-130J Israeli aircraft.

The Kuwait Air Force signed a contract for three KC-130J air refueling tankers in May 2010, with deliveries to begin in late 2013. The KC-130Js will extend the range of its F-18s and augment its fleet of three militarized L-100s.

Oman increased its C-130J order in August 2010 by adding two C-130Js to the single C-130J-30 ordered in 2009. Deliveries are to be completed by early 2014 or before.

The Mexican Government has requested 2 C-130J-30s.

RAAF C-130J-30 at Point Cook, 2006

RAF Hercules C4 (C-130J-30) in 2004

Two USMC KC-130Js of VMGR-352 during a training exercise

USAF C-130J-30 taxis to the runway at RIAT 2010

Lockheed L-100 Hercules

The Lockheed L-100 Hercules is the civilian variant of the prolific C-130 military transport aircraft made by the Lockheed Corporation. Its first flight occurred in 1964. Longer L-100-20 and L-100-30 versions were developed. L-100 production ended in 1992 with 114 aircraft delivered.

A Tepper Aviation L-382 taking off from Mojave Spaceport, California

In 1959, Pan American ordered 12 GL-207 Super Hercules to be delivered by 1962, it was to be powered by four 6,000 eshp Allison T61 turboprops. It was to be 23 ft 4 in (7.11 m) longer than the C-130B, a variant powered by 6,445 Rolls-Royce Tynes and a jet-powered variant with four Pratt & Whitney JT3D-11 turbofans were also under development. Both Pan American and Slick Airways (who had ordered six) cancelled their orders and the other variants did not evolve past design studies.

Lockheed decided to produce a commercial variant based on a de-militarised version of the C-130E Hercules. The prototype L-100 (N1130E) first flew on the 20 April 1964 when it carried out a 1 hour 25 minute flight. The type certificate was awarded on 16 February 1965. Twenty-one production aircraft were then built with the first delivery to Continental Air Services on 30 September 1965.

Slow sales led to the development of two new, longer versions, the L-100-20 and L-100-30, both of which were larger and more economical than the original model. Deliveries totaled 114 aircraft, with production ending in 1992.

An updated civilian version of the Lockheed Martin C-130J-30 Super Hercules was under development, but the program was placed on hold indefinitely in 2000 to focus on military development and production.

Northwest Territorial Airways L-100-30 at London Stansted Airport

Lockheed's L-100 freighters are the civil equivalents of the venerable military C-130 Hercules, and have proven to be of great utility, particularly in undeveloped countries.

Lockheed initiated design of the Hercules in response to a 1951 US Air Force requirement for a turboprop powered freighter. This resulted in the C-130 Hercules, which first flew in prototype form on August 23 1954. Design features included the high mounted wing, four Allison 501/T56 turboprops and the rear loading freight ramp. The USAF ordered the C-130 into series production in September 1952, and since that time more than 2500 have been built.

L-382 of Tepper Aviation at Mojave Spaceport

French L-100 in 1981

Saudi L-100 in 2011

The C-130's appeal to freight operators led Lockheed to develop a civil version. The first commercial versions were based on the C-130E model, and a demilitarised demonstrator first flew in April 1964. This initial civil development, the L-100 (L-382), was awarded civil certification in February 1965. This model was soon followed up by the series L-100 (L-382B), which introduced an improved freight handling system.

Sales of these initial versions were slow, leading Lockheed to develop the 2.54m (8ft 4in) stretched L-100-20 (L-382E), which offered better freight capacity and operating economics. The L-100-20 was certificated in October 1968, but was soon followed by the even longer L-100-30 (L-382G). The -30 was 2.03m (6ft 8in) longer than the -20, first flew in August 1970, and was delivered from December that year. Most civil Hercules sales have been of the L-100-30 variant. Although basically a civil aircraft, several L-100s are in service with military operators, e.g. in Algeria, Gabon and Kuwait. The last L-100 was built in 1992, while the last military Allison 501/T56 powered C-130 was delivered in 1996.

The L-100J would be a commercial derivative of the new generation C-130J Hercules II. Improvements would include new 3425kW (4591shp) Rolls-Royce (Allison) AE-2100D3 advanced turboprop engines driving six blade props, two crew EFIS flightdeck and significantly lower maintenance and operating costs. The C-130J first flew on April 5 1996, while US FAA civil certification was awarded in September 1998. The L-100J would be based on the stretched fuselage C-130J-30, but in 2000 the program was frozen as Lockheed martin focussed on the military variants.

Lockheed C-141 Starlifter

The Lockheed C-141 Starlifter was a military strategic airlifter in service with the Air Mobility Command (AMC) of the United States Air Force (USAF). The aircraft also served with AMC-gained airlift wings and air mobility wings of the Air Force Reserve Command (AFRC) and the Air National Guard (ANG) and, in later years, one air mobility wing of the Air Education and Training Command (AETC) dedicated to C-141, C-5, C-17 and KC-135 training.

Introduced to replace slower piston-engined cargo planes such as the C-124 Globemaster II, the C-141 was designed to requirements set in 1960 and first flew in 1963. Production deliveries of an eventual 285 planes began in 1965: 284 for the Air Force, and one for the National Aeronautics and Space Administration (NASA) for use as an airborne observatory. The aircraft remained in service for over 40 years until the USAF withdrew the last C-141s from service in 2006, after replacing the airlifter with the C-17 Globemaster III.

A United States Air Force C-141 C of the 452d Air Mobility Wing in 2003

In the early 1960s, the United States Air Force Military Air Transport Service (MATS) relied on a substantial number of propeller-driven aircraft for strategic airlift. As these aircraft were mostly obsolescent designs and the Air Force needed the benefits of jet power, the USAF ordered 48 Boeing C-135 Stratolifters as an interim step. The C-135 was a useful stop-gap, but only had side-loading doors and most bulky and oversize equipment would not fit, especially that employed by the U.S. Army.

In the spring of 1960 the Air Force released Specific Operational Requirement 182, calling for a new aircraft that would be capable of performing both strategic and tactical airlift missions. The strategic role demanded that the aircraft be capable of missions with a radius of at least 3,500 nmi (4,000 mi, 6,500 km) with a 60,000 lb (27,000 kg) load. The tactical role required it to be able to perform low-altitude air drops of supplies, and carry and drop paratroops in combat. Several companies responded to SOR 182, including Boeing, Lockheed and General Dynamics.

Lockheed responded to the requirement with a unique design: the Lockheed Model 300, the first large jet designed from the start to carry freight. The Model 300 had a swept high-mounted wing with four 21,000 lbf (93.4 kN) thrust TF33 turbofan engines pod-mounted below the wings.

An important aspect was the cabin floor's height of only 50 in (1.27 m) above the ground, allowing easy access to the cabin through the rear doors. The two rear side doors were designed to allow the aircraft to drop paratroopers (in August 1965 the type performed the first paratroop drop from a jet-powered aircraft). The rear cargo doors could be opened in flight to allow airborne freight drops.

The shoulder-mounted wings gave internal clearance in the cargo hold of 10 ft (3.05 m) wide, 9 ft (2.74 m) high and 70 ft (21.34 m) long. The size enabled the Starlifter to carry, for example, a complete LGM-30 Minuteman intercontinental ballistic missile in its container. The aircraft was capable of carrying a maximum of 70,847 lb (32,136 kg) over short distances, and up to 92,000 lb (41,730 kg) in the version configured to carry the Minuteman, which stripped other equipment. The aircraft could also carry up to 154 troops, 123 paratroopers or 80 litter patients.

Early C-141As of 436th Airlift Wing, MAC, at Brisbane Airport, Australia supporting the visit of President Lyndon B. Johnson, 22 October 1966.

The Apollo 11 Mobile Quarantine Facility is unloaded from a C-141 at Ellington Air Force Base, July 27, 1969.

President John F. Kennedy's first official act after his inauguration was to order the development of the Lockheed 300 on 13 March 1961, with a contract for five aircraft for test and evaluation to be designated the C-141.

One unusual aspect of the aircraft was that it was designed to meet both military and civil airworthiness standards. The prototype C-141A serial number 61-2775 was manufactured and assembled in record time, being rolled out of the Lockheed factory at Marietta, Georgia on 22 August 1963 and first flying on 17 December, the 60th anniversary of the Wright brothers' first flight.

The company and the Air Force then started an operational testing program and the delivery of 284 aircraft, initially to units of the MATS, later renamed the Military Airlift Command (MAC) in 1966.

An effort to sell the aircraft on the civilian market resulted in provisional orders from Flying Tiger Line and Slick Airways for four aircraft each.

These were to be a stretched version, 37 ft (11.28 m) longer than the C-141A, and marketed as the L-300 SuperstarLifter. Other changes were also incorporated to make more commercial including a different yoke.

The development was not sustained and only one civilian demonstration aircraft was built. When no commercial sales were made Lockheed donated the aircraft to NASA.

The prototype and development aircraft then began an intensive operational testing program including the first delivery to MATS (63-8078) on 19 October 1964 to the 1707th Air Transport Wing, Heavy (Training), Tinker Air Force Base, Oklahoma.

Testing continued and a Federal Aviation Authority type certificate was awarded on 29 January 1965.

The first delivery to an operational unit (63-8088) was made on 23 April 1965 to the 44th Air Transport Squadron, 1501st Air Transport Wing, Travis Air Force Base, California. Although operational testing continued, due to the United States' military involvement in South Vietnam, the C-141 was soon employed in operational sorties to the combat zone.

On 8 January 1966, following the disestablishment of MATS, all C-141s were transferred to the newly established Military Airlift Command (MAC).

The first strategic airlift flight of Operation Desert Shield was flown by a MAC C-141 of the 437th Military Airlift Wing out of Charleston AFB, SC, on 7 August 1990. The C-141 proved to be a workhorse airlifter of Operations Desert Shield and Desert Storm, flying 159,462 tons of cargo and 93,126 passengers during 8,536 airlift missions.

On 1 June 1992, following the disestablishment of Military Airlift Command, all C-141s and the airlift wings to which they were assigned were transferred to the newly-established Air Mobility Command (AMC).

Air Force Reserve Command (AFRC) and Air National Guard (ANG) C-141s and units were also transferred to AMC.

On 16 September 2004, the C-141 left service with all active duty USAF units, being confined to Air Force Reserve and Air National Guard units for the remainder of its operational service life.

As of 25 September 2005, there were only eight C-141 aircraft still flying, all from the Air Force Reserve's 445th Airlift Wing (445 AW) at Wright-Patterson AFB. In 2004, 2005, and 2006, the C-141s assigned to the 445 AW participated in missions to Iraq and Afghanistan, mostly for the medical evacuation of wounded service members.

The last eight C-141s were officially retired in 2006.

In 1994 one of the aircraft at Wright-Patterson AFB was identified by its crew chief as the Hanoi Taxi (AF Serial Number 66-0177), the first aircraft to land in North Vietnam in 1973 for Operation Homecoming in the final days of the Vietnam War, to repatriate American POWs from North Vietnam.

In 2005, Hanoi Taxi and other aircraft were marshalled by the Air Force to provide evacuation for those seeking refuge from Hurricane Katrina.

This aircraft and others evacuated thousands of people, including the medical evacuation (MEDEVAC) of hundreds of ill and injured.

With the 5 May 2005 announcement of the retirement of these last eight C-141s, the Hanoi Taxi embarked on a series of flights, giving veterans, some of whom flew out of POW captivity in Vietnam in this aircraft, the opportunity to experience one more flight before retirement.

On 6 May 2006, the Hanoi Taxi piloted by Col. James F. Blackman and Col. Benjamin Johnson, landed for the last time and was received in a formal retirement ceremony at the National Museum of the United States Air Force, located at Wright-Patterson Air Force Base near Dayton.

C-141 participating in Operation Deep Freeze

The original Starlifter model, designated C-141A, could carry 154 passengers, 123 paratroopers or 80 litters for wounded with seating for 16.

A total of 284 A-models were built. The C-141A entered service in April 1965. It was soon discovered that the aircraft's volume capacity was relatively low in comparison to its lifting capacity; it generally ran out of physical space before it hit its weight limit.

The C-141A could carry ten standard 463L master pallets and had a total cargo capacity of 62,700 lb (28,900 kg). It could also carry specialized cargoes, such as the Minuteman missile.

NASA obtained Lockheed's C-141 demonstrator, designated L-300.

The airplane was modified to house the Kuiper Airborne Observatory telescope for use at very high altitudes. This NASA NC-141A is now in storage at NASA Ames Research Center, Moffett Federal Airfield, CA.

In service, the C-141 proved to "bulk out" before it "massed out", meaning that it often had additional lift capacity that went wasted because the cargo hold was too full.

To correct the perceived deficiencies of the original model and utilize the C-141 to the fullest of its capabilities, the entire fleet of 270 in-service C-141As were stretched, adding needed payload volume.

The conversion program took place between 1977 and 1982, with first delivery taking place in December 1979. These modified aircraft were designated C-141B. It was estimated that this stretching program was equivalent to buying 90 new aircraft, in terms of increased capacity. Also added was a boom receptacle for inflight refueling.

The fuselage was stretched by adding "plug" sections before and after the wings, lengthening the fuselage a total of 23 ft 4 in (7.11 m) and allowing the carriage of 103 litters for wounded, 13 standard pallets, 205 troops, 168 paratroopers, or an equivalent increase in other loads.

A lengthened C-141B in front of a C-141A

In 1994, a total of 13 C-141Bs were given SOLL II (Special Operations Low-Level II) modifications, which gave the aircraft a low-level night flying capability, enhanced navigation equipment, and improved defensive countermeasures. These aircraft were operated by AMC in conjunction with Air Force Special Operations Command (AFSOC).

A total of 63 C-141s were upgraded throughout the 1990s to C-141C configuration, with improved avionics and navigation systems, to keep them up to date. This variant introduced some of the first glass cockpit technology to the aircraft, as well as improving reliability by replacing some mechanical and electromechanical components with their electronic equivalents.

A C-141 Starlifter leaves a vapor trail over Antarctica

A MAC C-141 transports the remains of the crew from Space Shuttle Challenger's doomed last mission to Dover AFB, Delaware.

President John F. Kennedy's first official act after his inauguration was to order the development of an all-jet transport to extend the reach of the nation's military forces. The Air Force required that the new aircraft be capable of performing both strategic and tactical airlift missions.

This meant that it must have the capability to perform at low altitudes, airdrop troops and supplies anywhere in the world, and travel at least 3,500 nautical miles (6,482km) with a 60,000 pound (27,216kg) load. Designed by the Lockheed-Georgia Company, the C-141A Starlifter was the result.

The first prototype (#61-2775) flew on 17 December 1963, the 60th anniversary of the Wright brothers' first flight.

The first C-141A, delivered to Tinker AFB, Oklahoma in October 1964, began squadron operations with the Military Air Transport Service (MATS) in April 1965. Starlifters made flights almost daily to Southeast Asia, carrying troops, equipment and supplies, and returning patients to U.S. hospitals.

The last C-141A (#67-0166) was delivered in February 1968.

When it entered service with MATS, the Starlifter has provided the U.S. Air Force with a fast and capacious long-range jet transport with which it could replace the slow C-124 Globemaster II and narrow-cabin C-135 Stratolifter. Drawing heavily on experience with the smaller C-130 Hercules, the Starlifter featured a fuselage of similar cross-section, a rear ramp and loading assembly with two large clamshell doors that could be opened in flight for airdrops, rear paratroop doors on both sides, and landing gear housed in external fairings.

A high-set wing, swept 25 degrees, was adopted for high-speed cruise, with powerful flaps provided for good low-speed field performance.

The aircraft also featured a T-tail, four underwing TF33 turbofan engines, and integral wing fuel tanks.

The C-141A was originally delivered to MATS in a natural silver finish, before adopting the Military Airlift Command's (MAC) white and grey scheme.

In the mid-1980s, after all but four A-models were converted to Bs, this scheme gave way to the "European One" camouflage. During the 1990s, much of the Air Mobility Command transport fleet, which also included the C-5 Galaxy and C-17 Globemaster III, adopted the current "Proud Grey" scheme.

Throughout its career the C-141 Starlifter has been the workhorse of the air mobility fleet, flying regular supply missions around the world in addition to special requirements.

The latter have included disaster relief, evacuations, aid delivery and missions in support of combat operations. Perhaps the Starlifter's finest hour came in the second half of 1990, when the entire fleet was instrumental in transporting much of the equipment for Operations Desert Shield and Desert Storm.

C-141B

Not long after the C-141A entered service, it became obvious that its maximum payload of 70,847 pounds (32,136kg) — or 92,000 pounds (41,731kg) on aircraft configured to carry LGM-30 Minuteman ICBMs — was rarely achieved, the aircraft frequently ran out of cabin volume long before its maximum payload weight had been reached. This problem was resolved by "stretching" the aircraft's fuselage.

During the 1970s, the entire fleet of 270 aircraft (minus the four NC-141A aircraft used as aerial testbeds) were returned to Lockheed for modification.

This process consisted of lengthening the aircraft by 23 feet, 4 inches (7.16m), which increased cargo capacity by about one-third — 2,171 extra cubic feet (61.48 cubic meters) — and increased the maximum payload weight from 70,847 pounds (32,136kg) to 90,880 pounds (41,222kg). Lengthening of the aircraft had the same effect as increasing the number of aircraft by 30 percent.

At the same time, a universal air refueling receptacle, with the ability to transfer 23,592 gallons in about 26 minutes, means longer nonstop flights and fewer fuel stops at overseas bases during worldwide airlift missions.

This inflight refueling system is housed within a characteristic "humped" fairing above the flight deck. This new capability provided the Starlifter with true global airlift capacity.

The prototype C-141B made its first flight on 24 March 1977 and Lockheed completed the final B-model on 29 June 1982. Of the 285 C-141A Starlifters built, 270 were converted to B-models.

C-141C

The Air Force has recently approved an upgrade for some "low-time" Starlifters that serve with Air Force Reserve and Air Guard units. C-141C modifications aim to preserve the remaining force by making reliability, maintainability and capability improvements necessary for effective use through 2006. Sixty-three aircraft in the current C-141B fleet will undergo major modification.

Each will receive a "glass cockpit" which features an all-weather flight control system, a Global Positioning System (GPS), chaff/flare dispensers, and a new fuel quantity indicating system. These upgrades should be complete before 2001.

The C-141 Starlifter provide low-altitude delivery of paratroops and equipment, and high-altitude delivery of paratroops. It can also airdrop equipment and supplies using the Container Delivery System (CDS).

The C-141 was the first aircraft designed to be compatible with the 463L Material Handling System, which permits off-loading 68,000 pounds (30,844kg) of cargo, refueling and reloading a full load, all in less than an hour.

Of inestimable value to the U.S. Air Force is the Starlifter's sheer versatility. Like that of the C-130 Hercules, the C-141's main hold is fitted with tie-down points and floor cleats that allow it to be rapidly reconfigured for different missions.

Some passenger configurations include: 200 troops seated in canvas side-facing seats, or 166 troops in rear-facing airline-type seats, or 103 litter patients plus attendants, or a combination of passengers and cargo. Rollers in the aircraft floor allow for quick and easy cargo pallet loading.

A palletized lavatory and galley can be installed quickly to accommodate passengers, and when palletized cargo is not being carried, the rollers can be turned over to leave a smooth, flat surface for loading vehicles.

Although most heavy equipment is moved by the C-5 Galaxy, the Starlifter can carry a Sheridan tank, an AH-1 Cobra helicopter, or five HMMWV vehicles.

Thirteen standard cargo pallets (88" x 108" @ 10,000 pound (4,536kg) capacity) can be admitted, and other loads can include aircraft engines, food supplies, fuel drums or weapons. The C-141 has pressurized cabin and crew station.

In its aeromedical evacuation (medevac) role, the Starlifter can carry about 103 litter patients, 113 ambulatory patients or a combination of the two.

It provides rapid transfer of the sick and wounded from remote areas overseas to hospitals in the United States.

Lockheed C-5 Galaxy

The Lockheed C-5 Galaxy is a large military transport aircraft built by Lockheed. It provides the United States Air Force (USAF) with a heavy intercontinental-range strategic airlift capability, one that can carry outsize and oversize cargos, including all air-certifiable cargo. The Galaxy has many similarities to its smaller C-141 Starlifter predecessor, and the later C-17 Globemaster. The C-5 is among the largest military aircraft in the world.

The C-5 Galaxy had a complicated development; significant cost overruns were experienced and Lockheed suffered significant financial difficulties. Shortly after entering service, fractures in the wings of many aircraft were discovered and the C-5 fleet were restricted in capability until corrective work was conducted. The C-5M Super Galaxy is an upgraded version with new engines and modernized avionics designed to extend its service life beyond 2040.

The C-5 Galaxy has been operated by USAF since 1969. In that time, it has been used to support US military operations in all major contingencies including Vietnam, Iraq, Yugoslavia and Afghanistan; as well as in support of allies, such as Israel during the Yom Kippur War and NATO operations in the Gulf War. The C-5 has also been used to distribute humanitarian aid and disaster relief, and support the US Space Shuttle program run by NASA.

A United States Air Force C-5 in-flight

In 1961, several aircraft companies began studying heavy jet transport designs that would replace the Douglas C-133 Cargomaster and complement Lockheed C-141 Starlifters. In addition to higher overall performance, the United States Army wanted a transport aircraft with a larger cargo bay than the C-141, whose interior was too small to carry a variety of their outsized equipment.

These studies led to the "CX-4" design concept, but in 1962 the proposed six-engine design was rejected, because it was not viewed as a significant advance over the C-141. By late 1963, the next conceptual design was named CX-X. It was equipped with four engines, instead of six engines in the earlier CX-4 concept. The CX-X had a gross weight of 550,000 pounds (249,000 kg), a maximum payload of 180,000 lb (81,600 kg) and a speed of Mach 0.75 (500 mph/805 km/h).

The cargo compartment was 17.2 ft (5.24 m) wide by 13.5 feet (4.11 m) high and 100 ft (30.5 m) long with front and rear access doors. To provide required power and range with only four engines required a new engine with dramatically improved fuel efficiency.

The criteria were finalized and an official request for proposal was issued in April 1964 for the "Heavy Logistics System" (CX-HLS) (previously CX-X).

In May 1964, proposals for aircraft were received from Boeing, Douglas, General Dynamics, Lockheed, and Martin Marietta.

General Electric, Curtiss-Wright, and Pratt & Whitney submitted proposals for the engines.

After a down select, Boeing, Douglas and Lockheed were given one-year study contracts for the airframe, along with General Electric and Pratt & Whitney for the engines. All three of the designs shared a number of features; all three placed the cockpit well above the cargo area to allow for cargo loading through a nose door.

The Boeing and Douglas designs used a pod on the top of the fuselage containing the cockpit, while the Lockheed design extended the cockpit profile down the length of the fuselage, giving it an egg-shaped cross section.

All of the designs had swept wings, as well as front and rear cargo doors allowing simultaneous loading and unloading. Lockheed's design featured a T-tail, while the designs by Boeing and Douglas had conventional tails.

The Air Force considered Boeing's design better than that of Lockheed, although Lockheed's proposal was the lowest total cost bid. Lockheed was selected the winner in September 1965, then awarded a contract in December 1965. General Electric's TF-39 engine was selected in August 1965 to power the new transport plane.

At the time GE's engine concept was revolutionary, as all engines beforehand had a bypass ratio of less than two-to-one, while the TF-39 promised and would achieve a ratio of eight-to-one, which had the benefits of increased engine thrust and lower fuel consumption.

The first C-5A Galaxy (serial number 66-8303) was rolled out of the manufacturing plant in Marietta, Georgia, on 2 March 1968. On 30 June 1968, flight testing of the C-5A began with the first flight, flown by Leo Sullivan, with the call sign "eight-three-oh-three heavy". Flight tests revealed that the aircraft exhibited a higher drag divergence Mach number than predicted by wind tunnel data.

The maximum lift coefficient measured in flight with the flaps deflected 40 degree was higher than predicted (2.60 vs. 2.38), but was lower than predicted with the flaps deflected 25 degrees (2.31 vs. 2.38) and with the flaps retracted (1.45 vs. 1.52).

Aircraft weight was a serious issue during design and development. At the time of the first flight, the weight was below the guaranteed weight, but by the time of the delivery of the 9th aircraft, had exceeded guarantees. In July 1969, during a fuselage up bending test, the wing failed at 128% of limit load, which is below the requirement that it sustain 150% of limit load.

Changes were made to the wing, but in a later test, in July 1970, it failed at 125% of limit load. A passive load reduction system, involving up rigged ailerons was incorporated, but the maximum allowable payload was reduced from 220,000 pounds to 190,000 pounds.

At the time, it was predicted that there was a 90% probability that no more than 10% of the fleet of 79 airframes would reach their fatigue life of 19,000 hours without cracking of the wing.

The fourth C-5A Galaxy 66-8306 in 1980s European One color scheme

Cost overruns and technical problems of the C-5A were the subject of a congressional investigation in 1968 and 1969. The C-5 program has the dubious distinction of being the first development program with a one billion dollar overrun.

Due to the C-5's troubled development, the Department of Defense abandoned Total Package Procurement.

In 1969 Henry Durham raised concerns about the C-5 production process with Lockheed, his employer; subsequently Durham was transferred and subjected to abuse until he resigned.

The Government Accountability Office (GAO) substantiated some of his charges against Lockheed; later the American Ethical Union honored Durham with the Elliott-Black Award.

Upon completion of testing in December 1969, the first C-5A was transferred to the Transitional Training Unit at Altus Air Force Base, Oklahoma. Lockheed delivered the first operational Galaxy to the 437th Airlift Wing, Charleston Air Force Base, SC, in June 1970.

Due to higher than expected development costs, in 1970 there were public calls for the government to split the substantial losses that Lockheed were experiencing.

Production was nearly brought to a halt in 1971 due to Lockheed going through financial difficulties, partly down to the C-5 Galaxy's development but also a civilian jet liner, the Lockheed L-1011.

The U.S. government gave loans to Lockheed to keep the company operational.

In the early 1970s, NASA considered the C-5 for the Shuttle Carrier Aircraft role, to transport the Space Shuttle to Kennedy Space Center.

However, they rejected it in favor of the Boeing 747, in part due to the 747's low-wing design. In contrast, the Soviet Union chose to transport its shuttles using the high-winged An-225, which derives from the An-124, which is similar in design and function to the C-5.

During static and fatigue testing cracks were noticed in the wings of several aircraft, and as a consequence the C-5A fleet was restricted to 80% of maximum design loads.

To reduce wing loading, load alleviation systems were added to the aircraft. By 1980, payloads were restricted to as low as 50,000 lb (23,000 kg) for general cargo during peacetime operations.

A $1.5 billion program, known as H-Mod, to re-wing the 76 completed C-5As to restore full payload capability and service life began in 1976.

After design and testing of the new wing design, the C-5As received their new wings from 1980 to 1987.

During 1976, numerous cracks were also found in the fuselage along the upper fuselage on the centerline, aft of the refueling port, extending back to the wing.

The cracks required a redesign to the hydraulic system for the visor, the front cargo entry point.

In 1974, Iran, then holding good relations with the United States, offered $160 million to restart C-5 production to enable Iran to purchase aircraft for their own air force; in a similar climate as to their acquisition of F-14 Tomcat fighters. However no C-5 aircraft were ever ordered by Iran, as the prospect was firmly halted by the Iranian Revolution in 1979.

As part of President Ronald Reagan's military policy, funding was made available for expansion of the USAF's airlift capability; however as the C-17 program was still some years from completion a new version of the C-5, the C-5B, was approved by Congress in July 1982 for purchase instead.

The first C-5B was delivered to Altus Air Force Base in January 1986. In April 1989, the last of 50 C-5B aircraft was added to the 77 C-5As in the Air Force's airlift force structure.

The C-5B includes all C-5A improvements and numerous additional system modifications to improve reliability and maintainability.

In 1998, the Avionics Modernization Program (AMP) began upgrading the C-5's avionics to include a glass cockpit, navigation equipment, and a new autopilot system. Another part of the C-5 modernization effort is the Reliability Enhancement and Re-engining Program (RERP).

The program will mainly replace the engines with newer, more powerful ones. Three C-5s underwent RERP as a test with full production in May 2008.

A total of 52 C-5s are contracted to be modernized, consisting of 49 B-, two C- and one A-model aircraft through the Reliability Enhancement and Re-Engining Program (RERP).

Over 70 changes and upgrades are incorporated in the program, including the newer General Electric engines. Five C-5M Super Galaxies have been produced. The RERP upgrade program is to be completed in 2016.

The C-5 is a large high-wing cargo aircraft. It has a distinctive high T-tail, 25 degree wing sweep, and four TF39 turbofan engines mounted on pylons beneath the wings. The C-5 is similar in layout to its smaller predecessor, the C-141 Starlifter. The C-5 has 12 internal wing tanks and is equipped for aerial refueling. It has both nose and aft doors for "drive-through" loading and unloading of cargo.

It has an upper deck seating area for 73 passengers and two loadmasters. The passengers face the rear of the aircraft, rather than forward. Its takeoff and landing distances, at maximum gross weight, are 8,300 ft (2,500 m) and 4,900 ft (1,500 m) respectively. Its high flotation main landing gear has 28 wheels to share the weight. The rear main landing gear is steerable for a smaller turning radius and it rotates 90 degrees horizontally before it is retracted after takeoff. The "kneeling" landing gear system permits lowering of the parked aircraft so the cargo floor is at truck-bed height to facilitate vehicle loading and unloading.

The C-5 has a Malfunction Detection Analysis and Recording (MADAR) system to identify errors throughout the aircraft. Some Galaxies have a Low Pressure Pneumatic System (LPPS) that utilizes a turbo compressor driven by bleed air to provide up to 150 psi pressure for inflating the aircraft's tires.

One of the unique features of the aircraft is the crosswind landing system that allows the landing gear to be offset up to 20 degrees either side of centerline; when the main landing gear was down (MLG) all the other 28 wheels would be slaved to the MLG and driven by hydraulic actuators to the same offset.

The C-5 features a cargo compartment 121 ft (37 m) long, 13.5 ft (4.1 m) high, and 19 ft (5.8 m) wide, or just over 31,000 cu ft (880 m3). The compartment can accommodate up to 36 463L master pallets or a mix of palletized cargo and vehicles.

The cargo hold of the C-5 is a foot longer than the entire length of the first powered flight by the Wright Brothers at Kitty Hawk. The nose and aft doors open the full width and height of the cargo compartment to permit faster and easier loading. Ramps are full width at each end for loading double rows of vehicles.

The Galaxy is capable of carrying nearly every type of the Army's combat equipment, including bulky items such as the 74 short tons (67 t) armored vehicle launched bridge (AVLB), from the United States to any location on the globe. A C-5 is capable of transporting up to six Boeing AH-64 Apaches or five Bradley Fighting Vehicles.

The first C-5A was delivered to the USAF on 17 December 1969. Wings were built up in the early 1970s at Altus AFB, Oklahoma, Charleston AFB, Dover AFB, Delaware, and Travis AFB, California. The C-5's first mission was on 9 July 1970, in Southeast Asia during the Vietnam War. C-5s were used to transport equipment and troops, including Army tanks and even some small aircraft, throughout the later years of the US action in Vietnam. In the final weeks of the war, prior to the Fall of Saigon, several C-5s were involved in evacuation efforts; during one such mission a C-5A crashed while transporting a large number of orphans.

C-5s have also been used to deliver support and reinforce various US allies over the years. During the Yom Kippur war in 1973, multiple C-5s and C-141 Starlifters delivered critical supplies of ammunition, replacement weaponry and other forms of aid to Israel, the US effort was named as Operation Nickel Grass. The C-5 Galaxy's performance in Israel was such that the Pentagon began to consider further purchases. The C-5 was regularly made available to support American allies, such as the British-led peacekeeper initiative in Zimbabwe in 1979.

On 24 October 1974, the Space and Missile Systems Organization successfully conducted an Air Mobile Feasibility Test where a C-5A Galaxy aircraft air dropped a 86,000 lb Minuteman ICBM from 20,000 ft over the Pacific Ocean. The missile descended to 8,000 ft before its rocket engine fired. The 10-second engine burn carried the missile to 20,000 ft again before it dropped into the ocean. The test proved the feasibility of launching an intercontinental ballistic missile from the air. Operational deployment was discarded due to engineering and security difficulties, though the capability was used as a negotiating point in the Strategic Arms Limitation Talks.

The C-5 has been used for several unusual functions; during the development of the secretive stealth fighter, the Lockheed F-117 Nighthawk, Galaxies were often used to carry partly disassembled aircraft, leaving no exterior signs as to their cargo. It remains the largest aircraft to ever operate in the Antarctic; Williams Field near McMurdo Station is capable of handling C-5 aircraft, the first of which landed there in 1989. The C-5 Galaxy was a major supply asset in the 1991 international coalition operations against Iraq's invasion of Kuwait, known as the First Gulf War. C-5s have routinely delivered relief aid and humanitarian supplies to areas afflicted with natural disasters or crisis, multiple flights were made over Rwanda in 1994.

The wings on the C-5As were replaced during the 1980s to restore full design capability. The USAF took delivery of the first C-5B on 28 December 1985 and the final one in April 1989. The reliability of the C-5 fleet has been a continued issue throughout its lifetime, however the C-5M upgrade program seeks in part to address this issue. Their strategic airlift capacity has been a key logistical component of U.S. military operations in Afghanistan and Iraq; following an incident during Operation Iraqi Freedom where one C-5 was damaged by a projectile, the installation of defensive systems has become a stated priority.

The C-5 AMP and RERP modernization programs plan to raise mission-capable rate to a minimum goal of 75%. Over the next 40 years, the U.S. Air Force estimates the C-5M will save over $20 billion. The first C-5M conversion was completed on 16 May 2006; C-5Ms began test flights at Dobbins Air Reserve Base in June 2006. The USAF decided to convert remaining C-5Bs and C-5Cs into C-5Ms with avionics upgrades and re-engining in February 2008. The C-5As will receive only the avionics upgrades.

In response to Air Force motions towards the retirement of the C-5 Galaxy, Congress implemented legislation that placed set limits upon retirement plans for C-5A models in 2003.

By 2005, 14 C-5As were retired. One was sent to the Warner Robins Air Logistics Center (WR-ALC) for tear down and inspection to evaluate structural integrity and estimate the remaining life for the fleet. 13 C-5As were sent to the Air Force's Aerospace Maintenance and Regeneration Group (AMARG) for inspection of levels of corrosion and fatigue.

The U.S. Air Force began to receive refitted C-5M aircraft in December 2008; full production of C-5Ms began in the summer of 2009. In 2009, the Congressional ban on the retirement of C-5s was overturned. The Air Force seeks to retire one C-5A for each 10 new C-17s ordered. In October 2011, the 445th Airlift Wing based at Wright-Patterson Air Force Base retired or reassigned all of its remaining C-5s; it has since reequipped with C-17s. In early February 2012, it was announced that the remaining 27 C-5As at Kelly Field, Texas; Memphis, Tennessee; and Martinsburg, West Virginia would be retired in fiscal years 2013 and 2014. Kelly is to receive C-5Ms currently assigned to Westover, Massachusetts and the other two wings are scheduled to receive C-17s.

On 13 September 2009, a C-5M set 41 new records; flight data was submitted to the National Aeronautic Association for formal recognition. The C-5M had carried a payload of 176,610 lb (80,110 kg) to over 41,100 ft (12,500 m) in 23 minutes, 59 seconds. Additionally, 33 time to climb records at various payload classes were set, and the world record for greatest payload to 6,562 ft (2,000 m) was broken. The aircraft was in the category of 551,160 to 661,390 lb (250,000 to 300,000 kg) with a takeoff weight of 649,680 lb (294,690 kg) including payload, fuel, and other equipment.

A C-5 taking off

People in line to enter the 445th Airlift Wing's first C-5A Galaxy in 2005

AETC

A detail of the C-5's nose assembly raised for loading and unloading.

The flight deck from the C-5B crash at Dover AFB in April 2006 being loaded into another C-5

Lockheed F-117 Nighthawk

The Lockheed F-117 Nighthawk is a single-seat, twin-engine stealth ground-attack aircraft formerly operated by the United States Air Force (USAF).

Its first flight was in 1981, and it achieved initial operating capability status in October 1983. The F-117 was "acknowledged" and revealed to the world in November 1988.

A product of Lockheed Skunk Works and a development of the Have Blue technology demonstrator, it became the first operational aircraft initially designed around stealth technology.

The F-117A was widely publicized during the Persian Gulf War of 1991.

It was commonly called the "Stealth Fighter" although it was a ground-attack aircraft.

The Air Force retired the F-117 on 22 April 2008, primarily because of the fielding of the F-22 Raptor and the impending introduction of the F-35 Lightning II.

Sixty-four F-117s were built, 59 of which were production versions with five demonstrators/prototypes.

F-117 flying over mountains

In 1964, Pyotr Ufimtsev, a Soviet mathematician, published a seminal paper titled Method of Edge Waves in the Physical Theory of Diffraction in the journal of the Moscow Institute for Radio Engineering, in which he showed that the strength of a radar return is related to the edge configuration of an object, not its size.

Ufimtsev was extending theoretical work published by the German physicist Arnold Sommerfeld. Ufimtsev demonstrated that he could calculate the radar cross-section across a wing's surface and along its edge.

The obvious and logical conclusion was that even a large aircraft could be made stealthy by exploiting this principle.

However, the airplane's design would make it aerodynamically unstable, and the state of computer technology in the early 1960s could not provide the kinds of flight computers which allow aircraft such as the F-117, and B-2 Spirit to stay airborne.

However, by the 1970s, when Lockheed analyst Denys Overholser found Ufimtsev's paper, computers and software had advanced significantly, and the stage was set for the development of a stealthy airplane.

The F-117 was born after combat experience in the Vietnam War when increasingly sophisticated Soviet surface-to-air missiles (SAMs) downed heavy bombers.

It was a black project, an ultra-secret program for much of its life, until the late 1980s. The project began in 1975 with a model called the "Hopeless Diamond" (a wordplay on the Hope Diamond because of its appearance).

The following year, the Defense Advanced Research Projects Agency issued Lockheed Skunk Works a contract to build and test two Stealth Strike Fighters, under the code name "Have Blue".

These subscale aircraft incorporated jet engines of the Northrop T-38A, fly-by-wire systems of the F-16, landing gear of the A-10, and environmental systems of the C-130.

By bringing together existing technology and components, Lockheed built two demonstrators under budget, at $35 million for both aircraft, and in record time.

The maiden flight of the demonstrators occurred on 1 December 1977. Although both aircraft were lost during the demonstration program, test data proved positive.

The success of Have Blue led the government to increase funding for stealth technology.

Much of that increase was allocated towards the production of an operational stealth aircraft, the Lockheed F-117A, under the program code name "Senior Trend".

The decision to produce the F-117A was made on 1 November 1978, and a contract was awarded to Lockheed Advanced Development Projects, popularly known as the Skunk Works, in Burbank, California.

The program was led by Ben Rich, who called on Bill Schroeder, a Lockheed mathematician, and Denys Overholser, a computer scientist, to exploit Ufimtsev's work.

The three designed a computer program called "Echo", which made it possible to design an airplane with flat panels, called facets, which were arranged so as to scatter over 99% of a radar's signal energy "painting" the aircraft.

The first YF-117A, serial number 79-0780, made its maiden flight from Groom Lake, Nevada on 18 June 1981, only 31 months after the full-scale development decision.

The first production F-117A was delivered in 1982, and operational capability was achieved in October 1983.

The Air Force denied the existence of the aircraft until 1988, when a grainy photograph was released to the public.

In April 1990 two were flown into Nellis Air Force Base, Nevada, arriving during daylight and publicly displayed to a crowd of tens of thousands. Quote: "Shortly after I arrived in Las Vegas to work at the Dunes, the USAF announced a static display of two aircraft for a one day open house at Nellis AFB." Five Full Scale Development (FSD) aircraft were built, designated "YF-117A".

The last of 59 production F-117s were delivered on 3 July 1990.

As the Air Force has stated, "Streamlined management by Aeronautical Systems Center, Wright-Patterson AFB, Ohio, combined breakthrough stealth technology with concurrent development and production to rapidly field the aircraft; The F-117A program demonstrates that a stealth aircraft can be designed for reliability and maintainability."

The aircraft maintenance statistics are comparable to other tactical fighters of similar complexity. Logistically supported by Sacramento Air Logistics Center, McClellan AFB, California, the F-117A was kept at the forefront of technology through a planned weapon system improvement program located at USAF Plant 42 at Palmdale, California.

The operational aircraft had the official designation of "F-117A". Most modern U.S. military aircraft use post-1962 designations in which the designation "F" is usually an air-to-air fighter, "B" is usually a bomber, "A" is usually a ground-attack aircraft, etc. (Examples include the F-15, the B-2, and the A-6.) The F-117 is primarily a ground-attack aircraft so its "F" designation is inconsistent with the DoD system, but it is an inconsistency that has been repeatedly employed by the U.S. Air Force with several of its ground attack aircraft since the late 1950s, including the Republic F-105 Thunderchief and General Dynamics F-111 Aardvark.

The designation "F-117" seems to indicate that it was given an official designation prior to the 1962 U.S. Tri-Service Aircraft Designation System and could be considered numerically to be a part of the earlier "Century series" of fighters. The assumption prior to the revealing of the aircraft to the public was that it would likely receive the designation F-19 as that number had not been used. However there were no other aircraft to receive a "100" series number following the F-111. Soviet fighters obtained by the United States via various means under the Constant Peg program were given F-series numbers for their evaluation by U.S. pilots, and with the advent of the Teen Series fighters, most often Century Series designations.

As with other exotic military aircraft types flying in the southern Nevada area, such as captured fighters, an arbitrary radio call of "117" was assigned. This same radio call had been used by the enigmatic 4477th Test and Evaluation Squadron, also known as the "Red Hats" or "Red Eagles", that often had flown expatriated MiGs in the area, but there was no relationship to the call and the formal F-19 designation then being considered by the Air Force. Apparently, use of the "117" radio call became commonplace and when Lockheed released its first flight manual (i.e., the Air Force "dash one" manual for the aircraft), F-117A was the designation printed on the cover.

A televised documentary quoted a senior member of the F-117A development team as saying that the top-notch USAF fighter pilots required to fly the new aircraft were more easily attracted to an aircraft with an "F" designation for fighter, as opposed to a bomber ("B") or attack ("A") designation.

The F-117 is shaped to deflect radar signals and is about the size of an F-15 Eagle. The single-seat Nighthawk is powered by two non-afterburning General Electric F404 turbofan engines, and has quadruple-redundant fly-by-wire flight controls. It is air refuelable. To lower development costs, the avionics, fly-by-wire systems, and other parts are derived from the General Dynamics F-16 Fighting Falcon, McDonnell Douglas F/A-18 Hornet and McDonnell Douglas F-15E Strike Eagle. The parts were originally described as spares on budgets for these aircraft, to keep the F-117 project secret.

The F-117 Nighthawk has a radar signature of about 0.025 m2 (0.269 sq ft). Among the penalties for stealth are lower engine power thrust, due to losses in the inlet and outlet, a very low wing aspect ratio, and a high sweep angle (50°) needed to deflect incoming radar waves to the sides. With these design considerations and no afterburner, the F-117 is limited to subsonic speeds.

The F-117A is equipped with sophisticated navigation and attack systems integrated into a digital avionics suite. It carries no radar, which lowers emissions and cross-section. It navigates primarily by GPS and high-accuracy inertial navigation. Missions are coordinated by an automated planning system that can automatically perform all aspects of an attack mission, including weapons release.

Front view of an F-117

Targets are acquired by a thermal imaging infrared system, slaved to a laser that finds the range and designates targets for laser-guided bombs. The F-117A's split internal bay can carry 5,000 lb (2,300 kg) of ordnance. Typical weapons are a pair of GBU-10, GBU-12, or GBU-27 laser-guided bombs, two BLU-109 penetration bombs, or two Joint Direct Attack Munitions (JDAMs), a GPS/INS guided stand-off bomb.

The F-117A's faceted shape (made from 2-dimensional flat surfaces) resulted from the limitations of the 1970s-era computer technology used to calculate its radar cross-section. Later supercomputers made it possible for subsequent planes like the B-2 bomber to use curved surfaces while staying stealthy, through the use of far more computational resources to do the additional calculations needed.

During the program's early years, from 1984 to mid-1992, the F-117A fleet was based at Tonopah Test Range Airport, Nevada where it served under the 4450th Tactical Group. Because the F-117 was classified during this time, the 4450th Tactical Group was "officially" located at Nellis Air Force Base, Nevada and equipped with A-7 Corsair II aircraft.

The 4450th was absorbed by the 37th Tactical Fighter Wing in 1989. In 1992, the entire fleet was transferred to Holloman Air Force Base, New Mexico, where it was placed under the command of the 49th Fighter Wing. This move also eliminated the Key Air and American Trans Air contract flights to Tonopah, which flew 22,000 passenger trips on 300 flights from Nellis to Tonopah per month.

F-117 pilots called themselves "Bandits". Each of the 558 Air Force pilots who have flown the F-117 have a Bandit number, such as "Bandit 52", that indicates the sequential order of their first flight in the F-117.

The F-117 has been used several times in war. Its first mission was during the United States invasion of Panama in 1989. During that invasion two F-117A Nighthawks dropped two bombs on Rio Hato airfield.

During the Persian Gulf War in 1991, the F-117A flew approximately 1,300 sorties and scored direct hits on 1,600 high-value targets in Iraq over 6,905 flight hours. Only 2.5% of the American aircraft in Iraq were F-117s, yet they struck more than 40% of the strategic targets.

F-117As dropped over 2,000 tons of precision-guided munitions and struck their targets with over an 80% success rate.

"Although the 37th Tactical Fighter Wing Provisional and its 42 stealth fighters represented just 2.5 percent of all allied fighter and attack aircraft in the Persian Gulf, the F-117As were assigned against more than 31 percent of the strategic Iraqi military targets attacked during the first 24 hours of the air campaign.

It was one of very few U.S. or Coalition aircraft types to strike targets in downtown Baghdad. Its stealth characteristics protected it from the SAM and AAA-saturated environment of the city, and due both to its precision-guided weaponry and its avoidance of having to jettison ordnance in defensive maneuvers F-117 missions proved less likely to cause civilian casualties and collateral damage.

Following its move to Holloman AFB in 1992, the F-117A and the men and women of the 49th Fighter Wing were deployed to Southwest Asia on multiple occasions. On their first deployment, with the aid of aerial refuelling, pilots flew non-stop from Holloman to Kuwait, a flight of approximately 18.5 hours – a record for single-seat fighters that stands today.

The F-117 was subsequently used in Operation Allied Force in 1999, Operation Enduring Freedom in 2001, and Operation Iraqi Freedom in 2003.

F-117s in formation

Only one F-117 (AF ser. no. 82-0806) was lost to enemy action. It was shot down during a mission against the Army of Yugoslavia on 27 March 1999, during Operation Allied Force. About 8:15 pm local time, SA-3s were fired from about 8 miles (13 km) away, launched by a Yugoslav version of the Soviet Isayev S-125 "Neva" (NATO name SA-3 "Goa") anti-aircraft missile system. The launcher was run by elements of the 3rd Battalion of the 250th Air Defence Missile Brigade under the command of Colonel Zoltán Dani.

According to NATO Commander Wesley Clark and other NATO generals, Dani detected F-117s by operating his radars on unusually long wavelengths, making the aircraft visible for brief periods. According to Dani in a 2007 interview, his troops spotted the aircraft on radar when its bomb-bay doors opened, raising its radar signature.

Some sources state one of the missiles detonated by its proximity fuze near the F-117, though its pilot has subsequently stated he is '99.98% sure' that the strike was a direct hit.

Dani said he kept most of his missile sites intact by frequently moving them, and had spotters looking for F-117s and other NATO aircraft. He oversaw the modification of his targeting radar to improve its detection. After the explosion, the aircraft became uncontrollable, forcing the pilot to eject.

The pilot was recovered a short time later by a U.S. Marine Corps combat search and rescue team. Photos show that the aircraft struck the ground at low speed in an inverted position, and that the airframe remained relatively intact.

The Serbs invited Russian personnel to inspect the aircraft's remains, compromising the then 25-year-old U.S. stealth technology. The F-117's pilot was initially misidentified. While the name "Capt Ken 'Wiz' Dwelle" was painted on the canopy, it was revealed in 2007 that the pilot was Lt. Col. Dale Zelko.

Some American sources claim that a second F-117A was damaged during the same campaign, allegedly on 30 April; the aircraft returned to base, but it supposedly never flew again. The stealth technology from the downed F-117 may have been acquired by Russia and China.

Though advanced for its time, the F-117's stealthy faceted airframe required a large amount of maintenance and was eventually superseded by streamlined shapes produced with computer-aided design. The Air Force had once planned to retire the F-117 in 2011, but Program Budget Decision 720 (PBD 720), dated 28 December 2005, proposed retiring it by October 2008 to free up an estimated $1.07 billion to buy more F-22 Raptors. PBD 720 called for 10 F-117s to be retired in FY 2007 and the remaining 42 in FY 2008, stating that other Air Force planes and missiles could stealthily deliver precision payloads, including the B-2 Spirit, F-22 and JASSM.

In late 2006, the Air Force closed the F-117 formal training unit (FTU), and announced the retirement of the F-117. The first six aircraft to be retired made their last flight on 12 March 2007 after a ceremony at Holloman AFB to commemorate the aircraft's career. Brigadier General David Goldfein, commander of the 49th Fighter Wing, said at the ceremony, "With the launch of these great aircraft today, the circle comes to a close – their service to our nation's defense fulfilled, their mission accomplished and a job well done. We send them today to their final resting place – a home they are intimately familiar with – their first, and only, home outside of Holloman."

Unlike most other Air Force aircraft which are retired to Davis-Monthan AFB for scrapping, or dispersal to museums, most of the F-117s were retired to their original hangars at the Tonopah Test Range Airport. At Tonopah, their wings were removed and the aircraft are stored in their original climate-controlled hangars. The decommissioning occurred in eight phases, with the operational aircraft retired to Tonapah in seven waves beginning on 13 March 2007, and ending with the last wave's arrival on 22 April 2008. Four aircraft were kept flying beyond April by the 410th Flight Test Squadron at Palmdale for flight test. By the beginning of August, two were remaining. The last F-117 (AF ser. no. 86-0831) left Palmdale to fly to Tonopah on 11 August 2008. With the last aircraft retired, the 410th was inactivated in a ceremony on 1 August 2008.

Five aircraft were placed in museums including the first 4 YF-117As and some remains of the F-117 shot down over Serbia. Through 2009, one F-117 has been scrapped. F-117 AF ser. no. 79-0784 was scrapped at the Palmdale test facility on 26 April 2008. It was the last F-117 at Palmdale and was scrapped to test an effective method for destroying F-117 airframes.

The actual withdrawal of these UFO man-made, was obviously a great mistake!

Although officially retired, the F-117 fleet remains intact, and photos show the aircraft carefully mothballed. F-117s have been spotted flying in the Nellis Bombing Range as recently as 2010.

F117A aircraft from the 37th Tactical Fighter Wing at Langley AFB, VA, prior to being deployed to Saudi Arabia for Operation Desert Shield.

F-117N "Seahawk"

In the early 1990s, Lockheed proposed an upgraded, carrier capable variant of the F-117 dubbed the "Seahawk" to the U.S. Navy as an alternative to the canceled A/F-X program. The unsolicited proposal was received poorly by the Department of Defense, which had little interest in the single mission capabilities of such an aircraft, particularly as it would take money away from the Joint Advanced Strike Technology program, which evolved into the Joint Strike Fighter.

The new aircraft would have differed from the land-based F-117 in several ways, including the addition "of elevators, a bubble canopy, a less sharply swept wing and reconfigured tail". The "N" variant would also be re-engined to use General Electric F414 turbofans instead of the older General Electric F404s. Furthermore the aircraft would be optionally fitted with hardpoints, allowing for an additional 8,000 lb (3,600 kg) of payload, and a new ground attack radar with air-to-air capability. In that role the F-117N could carry AIM-120 AMRAAM air-to-air missiles.

After being rebuffed by the Navy, Lockheed submitted an updated proposal that included afterburning capability and a larger emphasis on the F-117N as a multi-mission aircraft, rather than just an attack aircraft. In efforts to boost interest, Lockheed also proposed an F-117B land-based variant that shared most of the F-117N capabilities. This variant was proposed to both the US Air Force and the Royal Air Force. This renewed F-117N proposal was also known as the A/F-117X. Neither the F-117N or the F-117B were purchased by any party.

F-117N "Seahawk"

Lockheed Martin F-22 Raptor

The Lockheed Martin/Boeing F-22 Raptor is a single-seat, twin-engine fifth-generation supermaneuverable fighter aircraft that uses stealth technology. It was designed primarily as an air superiority fighter, but has additional capabilities that include ground attack, electronic warfare, and signals intelligence roles. Lockheed Martin Aeronautics is the prime contractor and is responsible for the majority of the airframe, weapon systems and final assembly of the F-22. Program partner Boeing Defense, Space & Security provides the wings, aft fuselage, avionics integration, and training systems.

The aircraft was variously designated F-22 and F/A-22 during the years prior to formally entering USAF service in December 2005 as the F-22A. Despite a protracted and costly development period, the United States Air Force considers the F-22 a critical component of U.S. tactical air power, and claims that the aircraft is unmatched by any known or projected fighter.

Lockheed Martin claims that the Raptor's combination of stealth, speed, agility, precision and situational awareness, combined with air-to-air and air-to-ground combat capabilities, makes it the best overall fighter in the world today. Air Chief Marshal Angus Houston, former Chief of the Australian Defence Force, said in 2004 that the "F-22 will be the most outstanding fighter plane ever built."

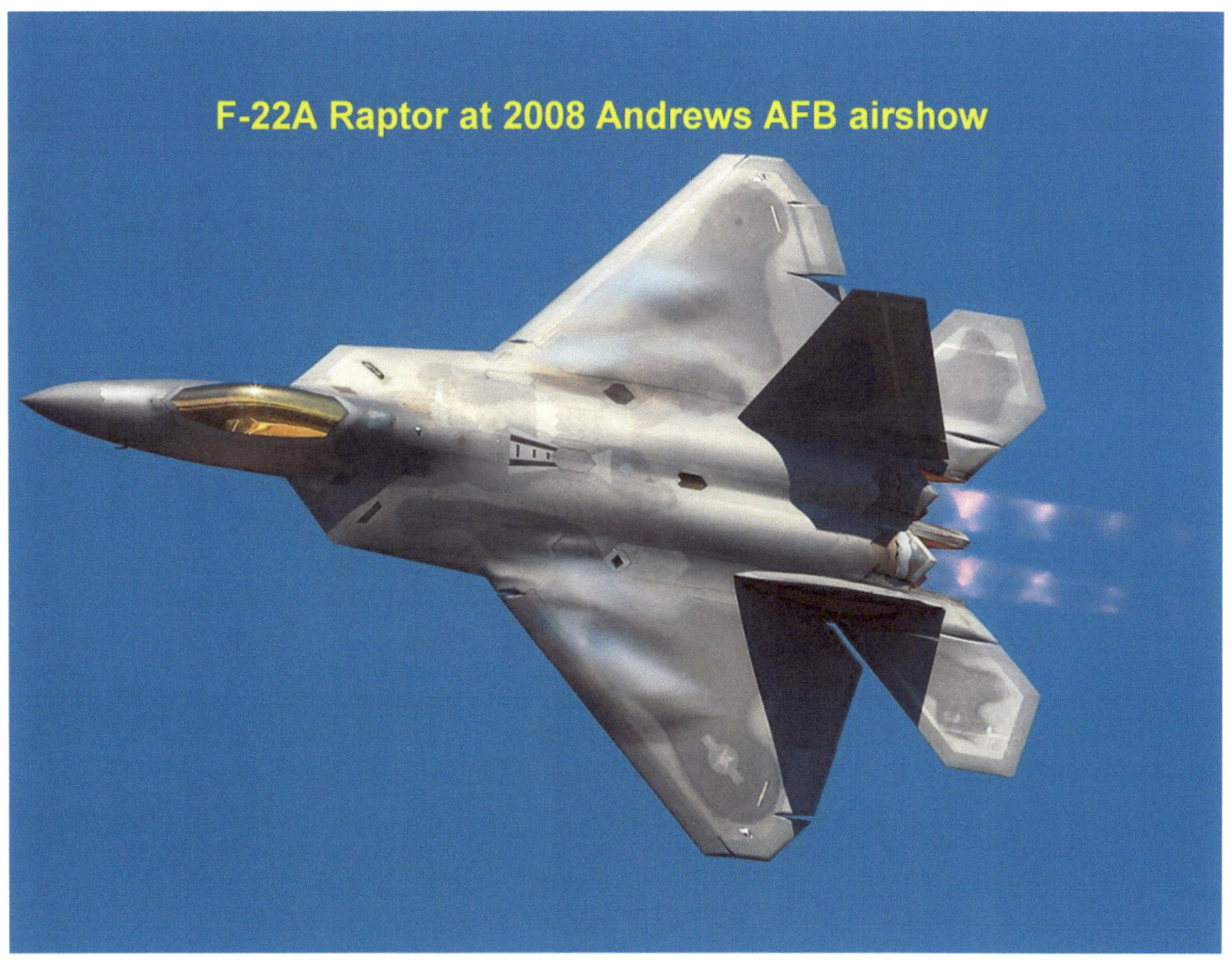

F-22A Raptor at 2008 Andrews AFB airshow

The high cost of the aircraft, a lack of clear air-to-air combat missions because of delays in the Russian and Chinese fifth-generation fighter programs, a U.S. ban on Raptor exports, and the ongoing development of the planned cheaper and more versatile F-35 resulted in calls to end F-22 production. In April 2009, the U.S. Department of Defense proposed to cease placing new orders, subject to Congressional approval, for a final procurement tally of 187 operational aircraft. The National Defense Authorization Act for Fiscal Year 2010 lacked funding for further F-22 production. The final F-22 rolled off the assembly line on 13 December 2011 during a ceremony at Dobbins Air Reserve Base.

Since 2010 the F-22 has been plagued by unresolved problems with its pilot oxygen systems which contributed to one crash and death of a pilot. In 2011 the fleet was grounded for four months before resuming flight operations, but reports of oxygen systems issues have continued. In July 2012, the Air Force announced that the hypoxia-like symptoms experienced were caused by a faulty valve in the pilots' pressure vest; the valve was replaced and changes to the filtration system were also made.

In 1981 the U.S. Air Force developed a requirement for an Advanced Tactical Fighter (ATF) as a new air superiority fighter to replace the F-15 Eagle and F-16 Fighting Falcon. This was influenced by the emerging worldwide threats, including development and proliferation of Soviet Su-27 "Flanker"- and MiG-29 "Fulcrum"-class fighter aircraft. It would take advantage of the new technologies in fighter design on the horizon including composite materials, lightweight alloys, advanced flight-control systems, more powerful propulsion systems, and stealth technology. A request for proposals (RFP) was issued in July 1986 and two contractor teams, Lockheed/Boeing/General Dynamics and Northrop/McDonnell Douglas, were selected on 31 October 1986 to undertake a 50-month demonstration phase, culminating in the flight test of two prototypes, the YF-22 and the YF-23.

Each design team produced two prototypes featuring one of two engine options, one featuring thrust vectoring. The Pratt & Whitney F119 turbofan with vectored thrust permits a tighter turning radius, a valuable capability in dogfights. The ATF's increasing weight and cost drove out some features during development. A dedicated infra-red search and track (IRST) system was downgraded from multi-color to single color and then deleted, the side-looking radars were deleted and the ejection seat requirement was downgraded from a fresh design to the existing McDonnell Douglas ACES II.

On 23 April 1991, the YF-22 was then announced by Secretary of the U.S. Air Force Donald Rice as the winner of the ATF competition. The YF-23 design was more stealthy and faster, but the YF-22 was more agile. The aviation press speculated that the YF-22 was also more adaptable to the Navy's Navalized Advanced Tactical Fighter (NATF), but the U.S. Navy abandoned NATF by 1992. In 1991, the air force planned to buy 650 aircraft.

The production F-22 model was unveiled on 9 April 1997 at Lockheed Georgia Co., Marietta, Georgia. It first flew on 7 September 1997. The first production F-22 was delivered to Nellis Air Force Base, Nevada, on 7 January 2003. In 2006 the Raptor's development team, composed of Lockheed Martin and over 1,000 other companies, plus the United States Air Force, won the Collier Trophy, American aviation's most prestigious award. In 2006, the USAF sought to acquire 381 F-22s, to be divided among seven active duty combat squadrons and three integrated Air Force Reserve Command and Air National Guard squadrons.

Several design changes were made from the YF-22 for production. The swept-back angle on the wing's leading edge was decreased from 48° to 42°, while the vertical stabilizer area was decreased by 20%. To improve pilot visibility, the canopy was moved forward 7 inches (178 mm), and the engine intakes were moved rearward 14 inches (356 mm).

The shape of the wing and stabilizer trailing edges was refined to improve aerodynamics, strength, and stealth characteristics. Also, the vertical stabilizer was shifted rearward.

During the development process the aircraft continued to gain weight at the cost of range and aerodynamic performance, even as capabilities were deleted or delayed in the name of affordability.

F-22 production was split up over many subcontractors across 46 states, in a strategy to increase Congressional support for the program. However the production split, along with the implementation of several new technologies were likely responsible for increased costs and delays. Many capabilities were deferred to post-service upgrades, reducing the initial cost but increasing total project cost. Each aircraft required "1,000 subcontractors and suppliers and 95,000 workers" to build. The F-22 was in production for 15 years, at a rate of roughly two per month.

The United States Air Force originally planned to order 750 ATFs at a cost of $26.2 billion, with production beginning in 1994; however, the 1990 Major Aircraft Review led by Defense Secretary Dick Cheney altered the plan to 648 aircraft beginning in 1996. The goal changed again in 1994, when it became 438 aircraft entering service in 2003 or 2004, but a 1997 Department of Defense report put the purchase at 339. In 2003, the Air Force said that the existing congressional cost cap limited the purchase to 277. In December 2004, the Department of Defense reduced procurement funding so only 183 aircraft could be bought. The Pentagon stated the reduction to 183 fighters would save $15 billion but raise the cost of each aircraft; this was implemented in the form of a multi-year procurement plan, which allowed for further orders later. The total cost of the program by 2006 was $62 billion.

In April 2006, the cost of the F-22 was assessed by the Government Accountability Office to be $361 million per aircraft. By April 2006, $28 billion had been invested in F-22 development and testing; while the Unit Procurement Cost was estimated at $177.6 million in 2006, based on a production run of 181 aircraft. It was estimated by the end of production, $34 billion will have been spent on procurement, resulting in a total program cost of $62 billion, around $339 million per aircraft. The incremental cost for an additional F-22 was estimated at about $138 million. In March 2012, the GAO increased the estimated cost to $412 million per aircraft.

On 31 July 2007, Lockheed Martin received a multi-year contract for 60 F-22s worth a total of $7.3 billion. The contract brought the number of F-22s on order to 183 and extended production through 2011. If production were restarted the cost for another 75 aircraft was estimated in 2009 to be an extra $70 million per unit.

No opportunity for export currently exists because the export sale of the F-22 is barred by American federal law. Current customers for U.S. fighters are either acquiring earlier designs such as the F-15, F-16 Fighting Falcon, and Boeing F/A-18E/F Super Hornet, or are waiting to acquire the Lockheed Martin F-35 Lightning II (Joint Strike Fighter), which contains technology from the F-22 but is designed to be cheaper, more flexible, and available for export. The F-35 will not be as agile as the F-22 or fly as high or as fast, but its radar and avionics will be more advanced. On 27 September 2006, Congress upheld the ban on foreign sales of the F-22; and confirmed this in December 2006.

The Japanese government showed interest in the F-22 for its Replacement-Fighter program. However, a sale would need approval from the Pentagon, State Department and Congress. It was stated that the F-22 would decrease the number of fighters needed by the Japan Air Self-Defense Force (JASDF), reducing engineering and staffing costs.

In August 2009, it was reported that the F-22 would require increases to the military budget beyond the historic 1 percent of GDP. In June 2009, Japanese Defense Minister Yasukazu Hamada said Japan still sought the F-22.

Some Australian politicians and defense commentators have proposed that Australia should purchase F-22s instead of the F-35. In 2006, Kim Beazley. leader of the Australian Labor Party supported this proposal on the grounds that the F-22 is a proven, highly capable aircraft, while the F-35 is still under development. However, Australia's Howard government ruled out purchase of the F-22, as its release for export is unlikely, and lacks sufficient ground/maritime strike capacity.

The following year, the newly-elected Rudd Government ordered a review of plans to procure the F-35 and F/A-18E/F Super Hornet, including an evaluation of the F-22's suitability.

The then Defence Minister Joel Fitzgibbon stated: "I intend to pursue American politicians for access to the Raptor". In February 2008, U.S. Defense Secretary Robert Gates said he had no objection to F-22 sales to Australia. However the RAAF found that the "F-22 Raptor cannot perform the strike or close air support roles planned for the JSF."

Thomas Crimmins of the Washington Institute for Near East Policy speculated in 2009 that the F-22 could be a strong diplomatic tool for Israel, strengthening the capability to strike Iranian nuclear facilities. Crimmins also stated the F-22 may be the only aircraft able to evade Russian S-300 air defense systems, which Russia may sell to Iran. However, Lockheed Martin has stated that the F-35 can handle the S-300, additionally Russia has stated they support and voted for United Nations sanctions on Iran preventing sales of the S-300.

The 2010 defense authorization bill included provisions that required the DoD to prepare a report on the costs and feasibility for an F-22 export variant and another report on the impact of F-22 export sales on the U.S. aerospace industry.

In 2006, David M. Walker, Comptroller General of the United States at the time, found that "the DOD has not demonstrated the need or value for making further investments in the F-22A program." During the two-month grounding of nearly 700 older F-15s in 2007, some U.S. Senators demanded Deputy Secretary of Defense Gordon England release three government reports supporting additional F-22s beyond the planned 183 jets. In December 2007, the USAF requested continued production beyond the planned 183 F-22s.

In January 2008, the Pentagon announced that it would ask Congress to fund additional F-22s to replace other aircraft lost in combat, and proposed that $497 million that would have been used to shut down the F-22 line be instead used to buy four extra F-22s, leaving the production line open beyond 2011 and allowing the next administration an option to buy more F-22s. Funds earmarked for line shutdown were redirected to repairs upon the F-15 fleet, delaying the end of F-22 production.

On 24 September 2008, Congress passed a defense spending bill funding continued production of the F-22. On 12 November 2008, the Pentagon released $50 million of the $140 million approved by Congress to buy parts for an additional four aircraft, thus leaving the Raptor program in the hands of the incoming Obama Administration.

On 6 April 2009, Secretary of Defense Gates called for the phasing out of F-22 production in fiscal year 2011, leaving the USAF with a production run of 187 fighters, minus losses. F-35 acquisition would be accelerated. On 17 June 2009 the House Armed Services Committee inserted $368.8 million in the budget towards a further 12 F-22s in FY 2011.

On 9 July 2009, General James Cartwright, Vice Chairman of the Joint Chiefs of Staff, explained to the U.S. Senate Committee on Armed Services his reasons for supporting termination of F-22 production. He stated that fifth-generation fighters need to be proliferated to all three services by shifting resources to the multirole F-35. He noted that commanders had concerns regarding electronic warfare (EW) capabilities, and that keeping the F/A-18 production line "hot" offered a fallback option to the F-35 in the EA-18G Growler.

On 21 July 2009, President Obama threatened to veto further F-22 production. On 21 July 2009, the Senate voted in favor of ending F-22 production. Secretary Gates said that the decision to end production was taken in light of the F-35's capabilities. On 29 July 2009, the Air National Guard's director asked for "60 to 70" F-22s for air sovereignty missions, noting that these could lack capabilities such as ground attack. On 30 July 2009, the House agreed to remove funds for an additional 12 aircraft and abide by the 187 cap. In mid-2010, Gates reduced the F-22 requirement from 243 to 187 aircraft, by lowering the preparations for two major regional conflicts to one. President Obama signed the National Defense Authorization Act for Fiscal Year 2010 in October 2009, without F-22 funding.

RAND estimated the cost of restarting production to build an additional 75 Raptors to be $17 billion or $227 million per aircraft. However Lockheed Martin has said that restarting the production line would only cost $200 million. The RAND paper was produced as part of an USAF study to determine the costs of retaining F-22 tooling for a future Service Life Extension Program (SLEP). The tooling for F-22 production will be documented in illustrated electronic manuals stored at the Sierra Army Depot.

Russian and Chinese fighter developments have fueled concern; General John Corley, head of Air Combat Command, wrote in a 2009 letter to a senator, "In my opinion, a fleet of 187 F-22s puts execution of our current national military strategy at high risk in the near- to mid-term". But Gates replied "Nonsense". On 8 January 2011, Gates clarified that Chinese fifth-generation fighter developments had been accounted when the number of F-22s was set, and that the United States would have a considerable advantage in stealth aircraft in 2025, even with F-35 delays. On 11 January 2011, China's J-20 stealth aircraft made its first flight, leading to speculation on the reactivation of F-22 production; An August 2008 RAND study concluded the F-22 would only play a minor role in a conflict with China over Taiwan as nearby bases would be rapidly shutdown by medium-range ballistic missiles (MRBMs); and distant bases would rely upon vulnerable aerial refueling tankers.

In December 2011, the 195th and final F-22 was completed (out of 8 test and 187 combat aircraft produced).

Because of the limited production run, there are zero attrition reserve aircraft and extra care is given at the Hill Air Force Base F-22 maintenance center to keep the entire fleet operational. Lockheed's retained tooling will be used to produce additional parts, if needed.

On 5 January 2001, Raptor 4005 flew with the Block 3.0 software, which was the first combat-capable avionics version. In June 2009, Increment 3.1 was tested at Edwards Air Force Base. This provided a basic ground-attack capability through Synthetic Aperture Radar mapping, Electronic attack and the GBU-39 Small Diameter Bomb. The Increment 3.1 Modification Team with the 412th Test Wing received the Chief of Staff Team Excellence Award for upgrading 149 Raptors. The fleet upgrade should start at the end of 2011. An additional $808 million will be spent in 2013 to implement the 3.1 upgrade. The first upgraded aircraft were delivered in early 2012.

Increment 3.2 was to add an improved SDB capability, an automatic ground collision avoidance system for low level operations (no longer planned) and enable use of the AIM-9X

Sidewinder and AIM-120D AMRAAM missiles. However, a helmet mounted cueing system has been deferred by technical issues. Increment 3.2 was expected to be fielded in FY15,[110] possibly including the Multifunction Advanced Data Link (MADL).

In July 2009, the USAF announced the modification of three business jets with the interim Battlefield Airborne Communications Node (BACN) to allow communication between F-22s and other platforms until MADL is installed. In March 2010, the USAF accelerated software portions of the Increment 3.2 upgrades to be completed in FY 2013, other upgrades will be completed later. Upgrading the first 183 aircraft to the 3.2 upgrade is estimated to cost $8 billion. In May 2009, Gen. Norton A. Schwartz and Air Force Secretary Michael B. Donley gave testimony to Congress that this would be paid for through the early retirement of legacy fighters. A total of 249 fourth-generation fighters were retired during Fiscal Year 2010. On 16 September 2009, Gates said "Our commitment to this aircraft is underscored by the 6 and-a half billion dollars... to upgrade the existing F-22 fleet to be fully mission-capable."

The USAF opened the Raptor enhancement, development and integration (REDI) contract to other bidders in January 2011 with a total budget of $16 billion. On 18 November 2011, the upgrade contract with Lockheed Martin was increased by $1.4 billion to a maximum value of $7.4 billion. This increment opens the way for further upgrades in 2012. The $11.7 billion allocated for the planned upgrades to the 3.2B level (of which $5.5 billion has been spent) includes almost $2 billion for structural repairs and reliability issues, but does not include related infrastructure costs. One of the goals of the reliability costs is to raise the fleet availability rate from its current level of 55.5% to 70.6% by 2015. The 3.2C level upgrades will be bid out as a separate project.

Lockheed Martin has proposed upgrades to add capabilities from the newer F-35. Elements such as MADL are delayed until the F-35 program is completed to reduce risk. One upgrade from the F-35 is new high-durability stealth coatings to lower maintenance. The Ada software language was blamed for slow progress and increased costs on the program, leading to a reorganization in 2011. Increment 3.2A in 2014 focuses on electronic warfare, communications and identification. Increment 3.2B in 2017 will support the AIM-9X and AIM-120D missiles. Increment 3.2C in 2019 may migrate some avionics to an open platform, allowing features to be added by various companies. Lockheed Martin is working on upgrading the AN/AAR-56 Missile Launch Detector (MLD) to provide situational awareness and defensive Infrared Search and Track similar to the F-35's SAIRST.

The current upgrade schedule is:

Increment 3.1 now entering service adds capabilities for SDB, SAR, and electronic attack.

Update 4 in 2012 will add a rudimentary capability for the AIM-120D.

Increment 3.2A will be fielded in 2014 with Link 16 and electronic warfare improvements. (The Link 16 receive capability has been moved up to 2013.)

Also by mid-2014 an automatic backup oxygen supply will be installed on all aircraft.

Update 5 in 2015 will add an initial capability for the AIM-9X.

In 2016 the fleet will be upgraded to 36 Block 20 training aircraft and 149 Block 30/35 operational aircraft.

Increment 3.2B in 2018 will add full capability for the air to air missiles, and "significantly improved ground threat geolocation". This schedule has slipped seven years because of "requirements and funding instability". Because of this delay the upgrade will be applied to fielded aircraft that have already consumed a significant fraction of their useful airframe lifespan.

Increment 3.2C was renamed 3.3 and while it is still being defined, it will include air traffic control updates.

Features not currently planned for addition or upgrades include:

The previously planned side-mounted AESA radar arrays

Infrared search and track (IRST)

Helmet-mounted sight

Powered air to surface missiles, the GBU-53 Small Diameter Bomb II, or other systems capable of engaging moving ground targets.

Because of these limitations, the Raptor will be unable to use the off-bore site and lock-on after launch features of its missiles.

The Raptor were designed with a lifespan of 30 years and 8000 flight hours, but to achieve this goal required a $100 million "structures retrofit program". Investigations are being made for upgrades to extend their useful lives further. The F-22 is expected to eventually be replaced by the fighter from the Next Generation Air Dominance program.

While no definitive, single cause has been found for the frequent oxygen deprivation issues that have killed at least one pilot, the F-22 will be upgraded with a 10 pound backup oxygen system, software upgrades and oxygen sensors to allow the pilots to operate normally in spite of the problem. Due to frequent stand downs during the investigation, the F-22 fleet averaged less than eight flight hours per month over 2011.

The F-22 Raptor is a fifth generation fighter that is considered a fourth-generation stealth aircraft by the USAF. Its dual afterburning Pratt & Whitney F119-PW-100 turbofans incorporate pitch axis thrust vectoring, with a range of ±20 degrees. The maximum thrust is classified, though most sources place it at about 35,000 lbf (156 kN) per engine. Maximum speed, without external weapons, is estimated to be Mach 1.82 in super cruise mode, as demonstrated by General John P.

Jumper, former U.S. Air Force Chief of Staff, when his Raptor exceeded Mach 1.7 without afterburners on 13 January 2005. With afterburners, it is "greater than Mach 2.0" (greater than 1,317 mph, 2,120 km/h). Former Lockheed chief test pilot Paul Metz stated that the Raptor has a fixed inlet, as opposed to variable intake ramps, and that the F-22 has a greater climb rate than the F-15, despite the F-15's higher thrust-to-weight ratio of 1.2:1 (the F-22 has a ratio closer to 1:1). The U.S. Air Force claims that the Raptor cannot be matched by any known or projected fighter types, and Lockheed Martin claims: "the F-22 is the only aircraft that blends super cruise speed, super-agility, stealth and sensor fusion into a single air dominance platform."

The ability of airframes to withstand both stress and heat is a major design factor, thus the F-22 makes use of various materials. The use of internal weapons bays allows the aircraft to maintain a comparatively higher performance while carrying a heavy payload over many other aircraft due to a lack of drag from external stores. It is one of only a few aircraft that can super cruise or sustain supersonic flight without the use of afterburners, which consume vastly more fuel. The F-22 can intercept time-critical or rapidly moving targets that a subsonic aircraft would not have the speed to follow and an afterburner-dependent aircraft would lack fuel to reach.

The F-22 is highly maneuverable, at both supersonic and subsonic speeds. It is extremely departure-resistant, enabling it to remain controllable at extreme pilot inputs. The Raptor's thrust vectoring nozzles allow the aircraft to turn tightly, and perform extremely high alpha (angle of attack) maneuvers such as the Herbst maneuver (or J-turn), Pugachev's Cobra, and the Kulbit. The F-22 is also capable of maintaining a constant angle of attack of over 60°, yet still having some control of roll. During June 2006 exercises in Alaska, F-22 pilots demonstrated that cruise altitude has a significant effect on combat performance, and routinely attributed their altitude advantage as a major factor in achieving an unblemished kill ratio against other U.S. fighters and 4th/4.5th generation fighters.

Members of the Fighter Mafia have criticized the shortcomings of the F-22 as compared to their ideal of the F-16 in pilot visibility, available flight hours for pilot training, a focus on Beyond-visual-range missile combat that ignores air combat history, stealth that does not cover important long range search frequencies, and very limited range. During Red Flag 2012, German pilots flying Typhoons demonstrated the weaknesses of the Raptor in close in combat due to its size, weight, disadvantage in IR sensors, and lack of off-bore site weapons. The Eurofighters had a maneuverability advantage because they were flying in a stripped down configuration, but the F-22's equipment is built into the airframe for stealth. So the Raptor always carries the additional size and weight.

The F-22 has a unique combination of speed, altitude, agility, sensor fusion and stealth that all work together to increase its effectiveness. Altitude plus advanced active and passive electronic warfare systems allow the F-22 to spot targets for its own weapons at considerable ranges. Altitude plus speed increases the reach of the F-22's own weapons. Altitude naturally increases the range from ground based defenses, which increases the effectiveness of stealth, and when combined with speed reduces the time defensive systems have to react to the F-22's attacks.

The F-22's avionics include BAE Systems E&IS radar warning receiver (RWR) AN/ALR-94, AN/AAR 56 Infra-Red and Ultra-Violet MAWS (Missile Approach Warning System) and the Northrop Grumman AN/APG-77 Active Electronically Scanned Array (AESA) radar. The AN/ALR-94 is a passive receiver system to detect radar signals; composed of more than 30 antennas blended into the wings and fuselage that provide all around coverage. It was described by Tom Burbage, former F-22 program head at Lockheed Martin, as "the most technically complex piece of equipment on the aircraft." It has a greater range (250+ nmi) than the radar, allowing the F-22 to limit its own radar emissions to maximize stealth. As a target approaches, the receiver can cue the AN/APG-77 radar to track the target with a narrow beam, which can be as focused down to 2° by 2° in azimuth and elevation.

The AN/APG-77 radar, designed for air superiority and strike operations, features a low-observable, active-aperture, electronically-scanned array that can track multiple targets in any weather. The AN/APG-77 changes frequencies more than 1,000 times per second to lower interception probability. Additionally, radar emissions can be focused in an electronic-attack capability to overload enemy sensors.

The radar's information is processed by two Raytheon Common Integrated Processor (CIP)s. Each CIP can process 10.5 billion instructions per second and has 300 megabytes of memory. Information can be gathered from the radar and other onboard and offboard systems, filtered by the CIP, and offered in easy-to-digest ways on several cockpit displays, enabling the pilot to remain on top of complicated situations. The F-22s avionics software has some 1.7 million lines of code, the majority involving processing data from the radar.

The radar has an estimated range of 125–150 miles, though planned upgrades will allow a range of 250 miles (400 km) or more in narrow beams. In 2007, tests by Northrop Grumman, Lockheed Martin, and L-3 Communications enabled the AESA system of a Raptor to act like a WiFi access point, able to transmit data at 548 megabits per second and receive at gigabit speed; this is far faster than the Link 16 system used by U.S. and allied aircraft, which transfers data at just over 1 Mbit/s.

The F-22 has a threat detection and identification capability comparative with the RC-135 Rivet Joint. The F-22's stealth allows it to safely operate far closer to the battlefield, compensating for the reduced capability. The F-22 is capable of functioning as a "mini-AWACS", however the radar is less powerful than dedicated platforms such as the E-3 Sentry. The F-22 allows its pilot to designate targets for cooperating F-15s and F-16s, and determine whether two friendly aircraft are targeting the same aircraft. This radar system can sometimes identify targets "many times quicker than the AWACS". The radar is capable of high-bandwidth data transmission; conventional radio "chatter" can be reduced via these alternative means. The IEEE-1394B data bus developed for the F-22 was derived from the commercial IEEE-1394 "FireWire" bus system. Sensor fusion combines data from all onboard and off board sensors into a common view to prevent the pilot from being overwhelmed.

In a critical article former Navy Secretary John Lehman wrote "at least the F-22s are safe from cyber attack. No one in China knows how to program the '83 vintage IBM software that runs them." Former Secretary of the USAF Michael Wynne blamed the use of the DoD's Ada as a reason for cost overruns and schedule slippages on many major military projects, including the F-22 Raptor. The F-22 uses the INTEGRITY-178B operating system from Green Hills Software, which is also used on the F-35, several commercial airliners and the Orion Crew Exploration Vehicle.

Herbert J. Carlisle has said that the F-22 can data link with the Tomahawk (missile).

The F-22 features a glass cockpit with no analog flight instruments. The primary flight controls are a force-sensitive side-stick controller and a pair of throttles. The monochrome head-up display offers a wide field of view and serves as a primary flight instrument for the pilot; information is also displayed upon six color liquid crystal display (LCD) panels. The canopy's dimensions are approximately 140 inches long, 45 inches wide, and 27 inches tall (355 cm x 115 cm x 69 cm) and weights 360 pounds. In August 2006, the Air Force Packaging Technology Engineering Facility (AFPTEF) was tasked with the design of a new shipping and storage container for the fragile F-22 Canopy.

The Raptor has integrated radio functionality for communicating on standard frequencies, the signal processing systems are virtualized rather than a separated hardware module. Radio functions are inactive during the strictest emissions control protocols (EMCON level) to maintain stealth; at lower EMCON levels the pilot may use the radio at will. There has been several media reports on the F-22's inability to communicate with other aircraft and funding cuts on integrating the new data linking standard, MADL. Voice communication is possible, but not data transfer yet. However, the Joint Tactical Radio System (JTRS), the software-defined radio project, was cancelled in October 2011 (before delivery to the F-22).

The integrated control panel (ICP) is a keypad system for entering communications, navigation, and autopilot data. Two 3 in × 4 in (7.6 cm × 10 cm) up-front displays located around the ICP are used to display integrated caution advisory/warning data, communications, navigation and identification (CNI) data and also serve as the stand-by flight instrumentation group and fuel quantity indicator. The stand-by flight group displays an artificial horizon, for basic instrument meteorological conditions.

The 8 in × 8 in (20 cm × 20 cm) primary multi-function display (PMFD) is located under the ICP, and is used for navigation and situation assessment. Three 6.25 in × 6.25 in (15.9 cm × 15.9 cm) secondary multi-function displays are located around the PMFD for tactical information and stores management.

The ejection seat is a version of the ACES II (Advanced Concept Ejection Seat) commonly used in USAF aircraft, with a center-mounted ejection control. The F-22 has a complex life support system, components include the on-board oxygen generation system (OBOGS), protective pilot garments and a breathing regulator/anti-g valve controlling flow and pressure to the pilot's mask and garments. The protective garments are designed to protect against chemical/biological hazards and cold-water immersion, to counter g-forces and low pressure at high altitudes, and to provide thermal relief. It was developed under the Advanced Technology Anti-G Suit (ATAGS) project. Suspicions regarding the performance of the OBOGS and life support equipment have been raised by several crashes.

The USAF initially wanted the aircraft to use direct voice input (DVI) controls. This was finally judged too technically risky and was abandoned.

The Raptor has three internal weapons bays: a large bay on the bottom of the fuselage, and two smaller bays on the sides of the fuselage, aft of the engine intakes. It can carry six compressed-carriage medium range missiles in the center bay and one short range missile in each of the two side bays. Four of the medium range missiles can be replaced with two bomb racks that can each carry one medium-size bomb or four small diameter bombs. Carrying missiles and bombs internally maintains its stealth capability and maintains lower drag resulting in higher top speeds and longer combat ranges.

Launching missiles requires opening the weapons bay doors for less than a second, while the missiles are pushed clear of the airframe by hydraulic arms. This reduces the Raptor's chance of detection by enemy radar systems due to launched ordnance and also allows the F-22 to launch long range missiles while maintaining super cruise.

The F-22 can also carry air-to-surface weapons such as bombs with Joint Direct Attack Munition (JDAM) guidance and the Small-Diameter Bomb, but cannot self-designate for laser-guided weapons. Air-to-surface ordnance is limited to 2,000 lb (compared to 17,000 lb of F/A-18). The Raptor has an M61A2 Vulcan 20 mm cannon in the right wing root. The M61A2 carries 480 rounds; enough ammunition for approximately five seconds of sustained fire.

The opening for the cannon's firing barrel is covered by a door when not in use to maximize stealth. The F-22 has been able to close to gun range in training dogfights while avoiding detection. The cannon fire is tracked by the aircraft's radar and displayed on the pilot's head up display.

The Raptor's very high sustained cruise speed and operational altitude add significantly to the effective range of both air-to-air and air-to-surface munitions. This gives it a 40% greater employment range for air to air missiles than the F-35. The USAF plans to procure the AIM-120D AMRAAM, reported to have a 50% increase in range compared to the AIM-120C. While specific figures remain classified, it is expected that JDAMs employed by F-22s will have twice or more the effective range of munitions dropped by legacy platforms. In testing, a Raptor dropped a 1,000 lb (450 kg) unpowered, free-fall JDAM from 50,000 feet (15,000 m), while cruising at Mach 1.5, striking a moving target 24 miles (39 km) away.

While the F-22 typically carries its weapons internally, the wings include four hard points, each rated to handle 5,000 lb (2,300 kg). Each hard point has a pylon that can carry a detachable 600 gallon fuel tank or a launcher holding two air-air missiles.

However, the use of external stores has a detrimental effect on the F-22's stealth, maneuverability and speed. The two inner hard points are "plumbed" for external fuel tanks; the hard points can be jettisoned in flight so the fighter can maximize its stealth after exhausting external stores. A stealth ordnance pod and pylon is being developed to carry additional weapons internally.

The stealth of the F-22 is due to a combination of factors, including the overall shape of the aircraft, the use of radar absorbent material (RAM), and attention to detail such as hinges and pilot helmets that could provide a radar return. However, reduced radar cross section is one of five facets of presence reduction addressed in the designing of the F-22. The F-22 was designed to disguise its infrared emissions, reducing the threat of infrared homing ("heat seeking") surface-to-air or air-to-air missiles, including its flat thrust vectoring nozzles. The aircraft was designed to be less visible to the naked eye; radio, heat and noise emissions are equally controlled.

The F-22 reportedly relies less on maintenance-intensive radar absorbent coatings than previous stealth designs like the F-117. These materials are susceptible to adverse weather conditions. Unlike the B-2, which requires climate-controlled hangars, the F-22 can undergo repairs on the flight line or in a normal hangar. The F-22 features a Signature Assessment System which delivers warnings when the radar signature is degraded and has necessitated repair. The exact radar cross section (RCS) remains classified; however, in 2009 Lockheed Martin released information indicating it to have a RCS (from certain angles) of −40 dBsm − the equivalent radar reflection of a "steel marble". Effectively maintaining the stealth features can decrease the F-22's mission capable rate to 62–70%.

The effectiveness of the stealth characteristics is difficult to gauge. The RCS value is a restrictive measurement of the aircraft's frontal or side area from the perspective of a static radar. When an aircraft maneuvers it exposes a completely different set of angles and surface area, potentially increasing visibility. Furthermore, stealth contouring and radar absorbent materials are chiefly effective against high-frequency radars, usually found on other aircraft. Low-frequency radars, employed by weather radars and ground warning stations, are alleged to be less affected by stealth technologies and are thus more capable as detection platforms. Rebecca Grant states that while faint or fleeting radar contacts make defenders aware that a stealth aircraft is present, interception cannot be reliably vectored to attack the aircraft.

The F-22 also includes measures designed to minimize its detection by infrared, including special paint and active cooling of leading edges to deal with the heat buildup encountered during super cruise flight.

The YF-22 was originally given the unofficial name "Lightning II", after the World War II fighter P-38, by Lockheed, which persisted until the mid-1990s when the USAF officially named the aircraft "Raptor". The aircraft was also briefly dubbed "Super Star" and "Rapier". The F-35 later received the Lightning II name on 7 July 2006. In September 2002, Air Force leaders changed the Raptor's designation to F/A-22. The new designation, mimicking the Navy's McDonnell Douglas F/A-18 Hornet, was intended to highlight plans for a ground-attack capability amid intense debate over the relevance of expensive air-superiority jets. The F-22 designation was reinstated on 12 December 2005, when the aircraft entered service.

Flight testing of the F-22 began in 1997. Raptor 4001 was retired and sent to Wright-Patterson AFB to be fired at for testing the fighter's survivability. Usable parts of 4001 would be used to make a new F-22. Another engineering and manufacturing development (EMD) F-22 was also retired and likely to be sent to be rebuilt. A testing aircraft was converted to a maintenance trainer at Tyndall AFB.

In May 2006, a released report documented a problem with a forward titanium boom on the aircraft. The problem was caused by a manufacturing defect in the heat-treating, making the boom less ductile than specified and potentially shortening the lives of roughly the first 80 F-22s. Modifications were implemented to restore full life expectancy.

On 15 December 2005 the USAF announced that the Raptor had reached its Initial Operational Capability (IOC).

During Exercise Northern Edge in Alaska in June 2006, 12 F-22s of the 94th FS downed 108 adversaries with no losses in simulated combat exercises. In two weeks of exercises, the Raptor-led Blue Force amassed 241 kills against two losses in air-to-air combat; neither Blue Force loss was an F-22. Shortly after was Red Flag 07-1 in February 2007.

Fourteen F-22s of the 94th FS supported Blue Force strikes and undertook close air support sorties themselves. Against superior numbers of Red Force Aggressor F-15s and F-16s, 6–8 F-22s maintained air dominance throughout. No sorties were missed because of maintenance or other failures, and only one Raptor was judged lost against the opposing force's defeat. F-22s also provided airborne electronic surveillance.

While attempting its first overseas deployment to the Kadena Air Base in Okinawa, Japan, on 11 February 2007, six F-22s flying from Hickam AFB, Hawaii experienced multiple computer failures while crossing the International Date Line (or 180th meridian of longitude dependent on software programming).

The failures included navigation and communication. The fighters were able to return to Hawaii by following tanker aircraft. Within 48 hours, the error was resolved and the journey resumed. 90th Fighter Squadron performed the first F-22 NORAD interception of two Russian Tu-95MS 'Bear-H' bombers over Alaska, on 22 November 2007. Since then, F-22s have also escorted probing Tu-160 "Blackjack" strategic bombers.

On 12 December 2007, General John D.W. Corley, USAF, Commander of Air Combat Command, officially declared the F-22s of the integrated active duty 1st Fighter Wing and Virginia Air National Guard 192d Fighter Wing fully operational, three years after the first Raptor arrived at Langley Air Force Base, Virginia. This was followed from 13 to 19 April 2008 by an Operational Readiness Inspection (ORI) of the integrated wing which rated it "excellent" in all categories, with a simulated kill-ratio of 221–0. The first pair of Raptors assigned to the 49th Fighter Wing became operational at Holloman Air Force Base, New Mexico, on 2 June 2008.

In December 2007, Secretary of the Air Force Michael Wynne requested that the F-22 be deployed to the Middle East; Secretary of Defense Gates rejected this option. Time suggested part of the reason for it not being used in the 2011 military intervention in Libya may have been its high unit cost.

On 28 August 2008, an unmodified F-22 from the 411th Flight Test Squadron performed in the first ever air-to-air refueling of an aircraft using synthetic jet fuel. The test was a part of a wider USAF effort to qualify aircraft to use the fuel, a 50/50 mix of JP-8 and a Fischer-Tropsch process-produced, natural gas-based fuel. Then in 2011 a Raptor made a supersonic flight on a 50% mixture of biofuel derived from camelina.

In April 2012, the U.S. military deployed several F-22s to an allied base less than 200 miles from Iran. The Iranian defense minister called the deployment of stealth fighters to the UAE a security threat.

In 2004 the F-22 had a mission ready rate of 62% fleet wide but this was improved to 70% in 2009, with the mission ready rate being predicted to reach 85% as the fleet reached 100,000 flight hours. In the early years of its service, the F-22 required more than 30 hours of maintenance for every flight hour, with the total cost per flight hour of $44,000, however in 2008 this figure had been lowered to 18.1 hours, and 10.5 hours by 2009; this is compared to the original Pentagon requirement of 12 maintenance hours per flight hour. At introduction the F-22 also had a Mean Time Between Maintenance (MTBM) of 1.7 hours, however in the 7 years since this has been improved to 3.2 hours, exceeding the original requirement of 3.0 hours by 2010.

Each Raptor requires a month-long packaged maintenance plan (PMP) after every 300 flight hours. The aircraft's stealth system, including its radar absorbing metallic skin, account for almost one third of all maintenance. Another source of maintenance problems is that many components require custom hand-fitting and are not interchangeable. The canopy of the aircraft was required to have a life of 800 hours, however the original design failed to meet this, averaging at 331 hours. In response to this the canopy was redesigned, and the new canopy has met its requirements of a 800 hour life expectancy.

In January 2007, it was reported that the F-22 maintained a 97% sortie rate (flying 102 out of 105 tasked sorties) while amassing a 144-to-zero kill ratio during "Northern Edge" air-to-air exercises held in Alaska, the first large-scale exercise in which the Raptor participated. Lt. Col. Wade Tolliver, the squadron commander of the 27th FS commented: "the stealth coatings are not as fragile as they were in earlier stealth aircraft. It isn't damaged by a rain storm and it can stand the wear and tear of combat without degradation." However, rain has caused "shorts and failures in sophisticated electrical components" when the Raptors were briefly posted to Guam.

In its 2012 budget request the USAF cut F-22 flight training hours by one-third to reduce the operating costs of flying the aircraft. The F-22 will continue in its role as an airshow demonstration aircraft and will be the only USAF solo aircraft demonstrator in 2012.

The manpower required to maintain each F-22A's stealth coatings has increased as the aircraft have gotten older. Part of this work involves removing and reapplying the stealth coatings in order to maintain a consistent mold line or outside shape of the aircraft.

In 2005 a USAF expert panel called the Raptor Aero medical Working Group recommended changes to deal with breathing problems on the F-22. These changes were not implemented at the time, but are under consideration in 2012. The changes included the installation of a backup oxygen system that will now be installed. Different reasons have been offered for the omission, including weight and cost. Problems were designed into Raptor by the 1990s USAF, that had shed "much of its scientific, medical and engineering expertise" from the Cold War. These problems only surfaced once F-22 flights became common after 2008.

On 11 February 2007, twelve Raptors flying from Hawaii to Japan were forced to turn back due to a software glitch in the F-22s' on-board navigational computers. The fleet was briefly grounded in February 2010 due to corroded ejection seat rods.

In May 2011, the Raptor fleet was grounded following the November 2010 crash near Elmendorf Air Force Base, Alaska. The F-22 had been restricted to flying below 25,000 ft while the Honeywell oxygen generating system was inspected. After five incidents of pilots suffering from hypoxia and decompression, General William M. Fraser III of Air Combat Command grounded the F-22 fleet indefinitely on 3 May 2011. In June 2011, the investigation broadened across the life support systems, and aircraft deliveries were stopped.

During the investigation it was found that F-22 pilots suffered from a "Raptor Cough" and other breathing problems nine times more often than the pilots of other fighters.

Some of the polymers used on the F-22 are a significant health risk to personnel; technicians are required to wear protective equipment such as eye protection, respirators and gloves to work on the aircraft. These same materials have been suggested to be behind the "Raptor Cough" and other health problems suffered by F-22 pilots and ground crews. USAF Maj. Gen. Charles Lyon suggested that the ground crew problems could be from dehydration or hypoglycemia, and that pilots are to be told to improve eating and to stay well hydrated.

In September 2011, the F-22 returned to flight with added pilot safety equipment and careful monitoring of crew and aircraft, while the investigation continued. On 21 October 2011, Langley's F-22s were grounded after a suspected oxygen system problem; Joint Base Elmendorf-Richardson grounded their aircraft as well. All aircraft were cleared to fly again on 25 October.

In late October 2011, Lockheed Martin was awarded a $24M contract to find the cause of the breathing difficulty, as well as providing other sustainment functions. During the investigation various causes were investigated including poisoning by carbon monoxide from the engines while warming up the aircraft inside the hangars, other chemicals have been inhaled from the on-board oxygen generating system (OBOGS), including oil fumes and propane, other toxins that could enter the aircraft environmental control systems such as neurotoxic diisocyanates from the polyurethane adhesive used to glue stealth materials to the aircraft, high concentrations of oxygen caused by high-G high altitude maneuvers that other American jet fighters can not duplicate, leading to the collapse of the pilot's pulmonary alveolus, and that the breathing system on board the Raptor might not provide sufficient oxygen in some situations.

In mid-December 2011, the Air Force said that there had been 14 episodes since September, when the F-22s returned to operation, in which pilots experienced "physiological incidents" that might have been caused by a lack of oxygen. Up to April 2012 seven serious accidents occurred with two pilots killed. Since the redeployment in September 2011, 11 incidents of pilots reporting hypoxia-like symptoms have been recorded.

Air Force pilots have reported being pressured to continue flying the aircraft in spite of fearing for their safety because of the still-unresolved problems with the oxygen system. And half of all F-22 pilots have "lost confidence in the aircraft". General Mike Hostage stated that some of the 200 Raptor pilots have asked to transfer to other areas because of the problem. Hostage said that he will start flying the aircraft himself to better understand the relevant issues. Hostage qualified on the Raptor in June 2012. But he soon stepped down from the cockpit after admitting that the limited flight hours available needed to be reserved for the actual pilots.

In May 2012, it was announced that two pilots, Major Jeremy Gordon and Captain Josh Wilson, who had appeared on the CBS news program 60 Minutes, saying they didn't feel safe in the jet, were considered whistleblowers protected by the federal whistleblower legislation. Defense Secretary Leon Panetta subsequently ordered that all F-22 flights stay "within the proximity of potential landing locations" as more pilots came forward to report hypoxia-like symptoms.

Another cause may be the pressure-garment worn by the pilots, which may interfere with breathing; the fix may be to use the more evolved design built for the F-35. In the meantime the pilots have been instructed to not wear the pressure vests during routine flights. The same vests have shown an almost "unanimous failure rate in testing".

The two most recent breathing incidents were determined to be the results of mechanical problems unrelated to the general problem.

In July 2012 the Pentagon concluded that a pressure valve on vests worn during high-altitude flights and a carbon air filter were the likely sources of at least some hypoxia-like symptoms, and that long distance flights can resume without the vests and therefore at lower altitudes. The carbon filters were also changed to a different model to reduce shedding of inert, but alarming, black dust into the pilot's lungs. The first overseas flight under the new rules was a routine rotation at Kadena Air Base, Okinawa.

On 1 August 2012 it was announced that the connector hoses and valves of the "Combat Edge" upper pressure garment were at fault in many recent hypoxia-like instances, specifically the breathing regulator/anti-G (BRAG) valve that is used to inflate the vest. Later that month the flight duration restrictions were lifted, but without an effective pressure vest the Raptors remained restricted to the same altitudes as other fighters. On 19 September 2012, USAF General Gilmary M. Hostage III stated that the main source of the problems was not the hardware, but the "human physiology" of the pilots. He added that F-22 pilots would undergo training to teach them how to react appropriately to oxygen issues while operating the jet. One of the changes would be to revert to automatic rather than the maximum setting for oxygen as the USAF prepared to open up the Raptors and update the firmware on the OBOGS along with the installation of the backup system. A further investigation found that the OBOGS unexpectedly reduced oxygen levels when the aircraft performed high-G maneuvers.

F-22 with drop tanks in transit to Kadena Air Base, Japan from Langley Air Force Base, Virginia

The same valve had earlier been tried on the F-15 and F-16 aircraft, but found defective and unnecessary. When the higher attitudes of the F-22 then required pressurization of the pilot's vest, the same faulty valve was reintroduced without fixes.

In late 2012, Lockheed was awarded contracts to install an automatic oxygen backup system in addition to the primary system and manual backup.

The USAF is also considering the installation of equipment to determine if the pilots brains are functioning. With a grant from the USAF a pilot helmet sized EEG scanner has been developed by the University of California San Diego together with the Taiwanese National Chiao-Tung University and Taiwanese chipmakers.

An F-22, based at Elmendorf AFB, Alaska, over mountain terrain

Lockheed Martin F-35 Lightning II

The Lockheed Martin F-35 Lightning II is a family of single-seat, single-engine, fifth generation multirole fighters under development to perform ground attack, reconnaissance, and air defense missions with stealth capability.

The F-35 has three main models; the F-35A is a conventional takeoff and landing variant, the F-35B is a short take off and vertical-landing variant, and the F-35C is a carrier-based variant.

The F-35 is descended from the X-35, the product of the Joint Strike Fighter (JSF) program. JSF development is being principally funded by the United States, with the United Kingdom and other partner governments providing additional funding.

The partner nations are either NATO members or close U.S. allies. It is being designed and built by an aerospace industry team led by Lockheed Martin. The F-35 carried out its first flight on 15 December 2006.

The United States plans to buy a total of 2,443 aircraft to provide the bulk of its tactical airpower for the U.S. Air Force, Marine Corps and Navy over the coming decades. The United Kingdom, Italy, Netherlands, Australia, Canada, Norway, Denmark, Turkey, Israel and Japan are part of the development program and may equip their air services with the F-35.

An F-35C Lightning II, marked CF-1, conducts a test flight over the Chesapeake Bay

The JSF program was designed to replace the United States military F-16, A-10, F/A-18 (excluding newer E/F "Super Hornet" variants) and AV-8B tactical fighter aircraft. To keep development, production, and operating costs down, a common design was planned in three variants that share 80 percent of their parts:

F-35A, conventional take off and landing (CTOL) variant.

F-35B, short-take off and vertical-landing (STOVL) variant.

F-35C, carrier-based CATOBAR (CV) variant.

George Standridge of Lockheed Martin has said that the F-35 will be four times more effective than legacy fighters in air-to-air combat, eight times more effective than legacy fighters in air-to-ground combat, and three times more effective than legacy fighters in reconnaissance and suppression of air defenses – while having better range and requiring less logistics support and having around the same procurement costs (if development costs are ignored) as legacy fighters. Further, the design goals call for the F-35 to be the premier strike aircraft through 2040 and be second only to the F-22 Raptor in air superiority.

While the actual JSF development contract was signed on 16 November 1996, the contract for System Development and Demonstration (SDD) was awarded on 26 October 2001 to Lockheed Martin, whose X-35 beat the Boeing X-32.

Although both aircraft met or exceeded requirements, the X-35 design was considered to have less risk and more growth potential. The designation of the new fighter as "F-35" is out-of-sequence with standard DoD aircraft numbering, by which it should have been "F-24". It came as a surprise even to Lockheed, which had been referring to the aircraft in-house by this expected designation.

Based on wind tunnel testing, Lockheed Martin slightly enlarged its X-35 design into the F-35. The forward fuselage is 5 inches (130 mm) longer to make room for avionics. Correspondingly, the horizontal stabilizers were moved 2 inches (51 mm) rearward to retain balance and control. The top surface of the fuselage was raised by 1 inch (25 mm) along the center line. Also, it was decided to increase the size of the F-35B STOVL variant's weapons bay to be common with the other two variants. Manufacturing of parts for the first F-35 prototype airframe began in November 2003.

The F-35B STOVL variant was in danger of missing performance requirements in 2004 because it weighed too much; reportedly, by 2,200 lb (1,000 kg) or 8 percent. In response, Lockheed Martin added engine thrust and thinned airframe members; reduced the size of the common weapons bay and vertical stabilizers; re-routed some thrust from the roll-post outlets to the main nozzle; and redesigned the wing-mate joint, portions of the electrical system, and the portion of the aircraft immediately behind the cockpit.

Many of the changes were applied to all three variants to maintain high levels of commonality. By September 2004, the weight reduction effort had reduced the aircraft's design weight by 2,700 pounds (1,200 kg).

On 7 July 2006, the U.S. Air Force officially announced the name of the F-35: Lightning II, in honor of Lockheed's World War II-era twin-prop Lockheed P-38 Lightning and the Cold War-era jet, the English Electric Lightning. English Electric Company's aircraft division was a predecessor of F-35 partner BAE Systems. Lightning II was also an early company name for its fighter that was later named the F-22 Raptor.

On 19 December 2008, Lockheed Martin rolled out the first weight-optimized F-35A (designated AF-1). It is the first F-35 to be produced at a full-rate production speed and is structurally identical to the production F-35As that will be delivered starting in 2010.

Lockheed Martin Aeronautics is the prime contractor and performs aircraft final assembly, overall system integration, mission system, and provides forward fuselage, wings and flight controls system. Northrop Grumman provides Active Electronically Scanned Array (AESA) radar, electro-optical Distributed Aperture System (DAS), Communications, Navigation, Identification (CNI), center fuselage, weapons bay, and arrestor gear. BAE Systems provides aft fuselage and empennages, horizontal and vertical tails, crew life support and escape systems, Electronic warfare systems, fuel system, and Flight Control Software (FCS1). Alenia will perform final assembly for Italy and, according to an Alenia executive, assembly of all European aircraft with the exception of Turkey and the United Kingdom. The F-35 program has seen a great deal of investment in automated production facilities. For example, Handling Specialty produced the wing assembly platforms for Lockheed Martin. In November 2009, Jon Schreiber, head of F-35 international affairs program for the Pentagon, said that the U.S. will not share the software code for the F-35 with its allies.

As of 5 January 2009, six F-35s were complete, including AF-1 and AG-1, and 17 were in production. "Thirteen of the 17 in production are pre-production test aircraft, and all of those will be finished in 2009," said John R. Kent, acting manager of F-35 Lightning II Communications at Lockheed Martin Aeronautics Company. "The other four are the first production-model planes, and the first of those will be delivered in 2010 to the U.S. Air Force, and will go to Eglin Air Force Base." On 6 April 2009, U.S. Secretary of Defense Robert Gates proposed speeding up production for the U.S. to buy 2,443 F-35s.

In August 2010, Lockheed Martin announced delays in resolving a "wing-at-mate overlap" production problem, which would slow initial production.

In October 2011, two F-35B VTOL aircraft conducted three weeks of initial sea trials aboard USS Wasp (LHD-1), logging more than 28 hours of flight time including 72 short takeoffs and 72 vertical landings.

In 2006, the GAO warned that excessive concurrency in F-35 production and testing might result in expensive refits for several hundred aircraft planned to be produced before completion of tests. In November 2010, the GAO found that "Managing an extensive, still-maturing global network of suppliers adds another layer of complexity to producing aircraft efficiently and on-time" and that "due to the extensive amount of testing still to be completed, the program could be required to make alterations to its production processes, changes to its supplier base, and costly retrofits to produced and fielded aircraft, if problems are discovered." The United States Air Force (USAF) budget data in 2010, along with other sources, projects the F-35 to have a flyaway cost from US$89 million to US$200 million over the planned production of F-35s. In February 2011, the Pentagon put a price of $207.6 million for each of the 32 aircraft to be acquired in FY2012, rising to $304.15 million ($9,732.8/32) if its share of RDT&E spending is included.

In 2011, program head Vice Adm. David Venlet confirmed that the concurrency built into the program "was a miscalculation". This was during a contract dispute where the Pentagon insisted that Lockheed Martin help cover the costs of applying fixes found during testing to aircraft already produced. Lockheed Martin objected that the cost sharing posed an uninsurable unbounded risk that the company could not cover, and later responded that the "concurrency costs for F-35 continue to reduce".

However, the Senate Armed Services Committee strongly backed the Pentagon position. In December 2011, Lockheed Martin agreed to a cost sharing agreement. The Aerospace Industries Association trade group warned that such changes would force them to change their behavior and anticipate cost overruns in their future contract bids. As of 2012, problems found in flight testing are expected to continue to lead to elevated levels of engineering changes (to be made to newly produced aircraft and retrofitted onto previously produced aircraft) through 2019. The total additional cost for concurrency in the program is around $1.3 billion.

In 2012, General Norton A. Schwartz decried the "foolishness" of reliance on computer models to settle the final design of the aircraft before flight testing found the issues that needed redesign.

On 21 April 2009, media reports, citing Pentagon sources, said that during 2007 and 2008, computer spies had managed to copy and siphon off several terabytes of data related to the F-35's design and electronics systems, potentially enabling the development of defense systems against the aircraft. However, Lockheed Martin has rejected suggestions that the project has been compromised, saying that it "does not believe any classified information had been stolen".

However, other sources have suggested that the incident caused a redesign of the aircraft's hardware and software to be more resistant to cyber attack. BAE Systems was reported to be the target of the cyber espionage that may have stolen secrets related to the F-35.

On 9 November 2009, Ashton Carter, under-secretary of defense for acquisition, technology and logistics, acknowledged that the Pentagon "joint estimate team" (JET) had found possible future cost and schedule overruns in the project and that he would be holding meetings to attempt to avoid these. On 1 February 2010, Gates removed the JSF Program Manager, U.S. Marine Corps Major General David Heinz, and withheld $614 million in payments to Lockheed Martin because of program costs and delays.

On 11 March 2010, a report from the Government Accountability Office to United States Senate Committee on Armed Services projected the overall unit cost of an F-35A to be $112M in today's money. In 2010, Pentagon officials disclosed that the F-35 program has exceeded its original cost estimates by more than 50 percent. An internal Pentagon report critical of the JSF project states that "affordability is no longer embraced as a core pillar".

On 24 March, Gates termed the recent cost overruns and delays as "unacceptable" in a testimony before the U.S. Congress. He characterized previous cost and schedule estimates for the project as "overly rosy".

However, Gates insisted the F-35 would become "the backbone of U.S. air combat for the next generation" and informed the Congress that he had expanded the development period by an additional 13 months and budgeted $3 billion more for the testing program while slowing down production. Lockheed Martin expects to reduce government cost estimates by 20%.

In November 2010, as part of a cost-cutting measure, the co-chairs of the National Commission on Fiscal Responsibility and Reform suggested canceling procurement of the F-35B and halving orders of F-35As and F-35Cs. At the same time, Air Force Magazine reported that "Pentagon officials" are considering canceling the F-35B because its short range means that the bases or ships it operates from will be within range of hostile tactical ballistic missiles.

However, Lockheed Martin consultant Loren B. Thompson said that this rumor is merely a result of the usual tensions between the U.S. Navy and Marine Corps, and there is no alternative to the F-35B as an AV-8B replacement. He also confirmed that there would be further delays and cost increases in the development process because of technical problems with the aircraft and software, but blamed most of the delays and extra costs on redundant flight tests.

The Center for Defense Information estimated that the program would be restructured with an additional year of delay and $5 billion in additional costs. On 5 November 2010, the Block 1 software flew for the first time on BF-4 which included information fusion and initial weapons-release capability. As of the end of 2010, only 15% of the software remains to be written, but this includes the most difficult sections such as data fusion. But in 2011, it was revealed that only 50% of the eight million lines of code had actually been written and that it would take another six years and 110 additional software engineers in order to complete the software for this new schedule. The total estimated lines of code for the entire program (onboard and offboard) had grown from 15 million lines to 24 million lines by 2012.

In January 2011, Defense Secretary Robert Gates expressed the Pentagon's frustration with the skyrocketing costs of the F-35 program when he said "The culture of endless money that has taken hold must be replaced by a culture of restraint." Focusing his attention on the troubled F-35B, Gates ordered "a two-year probation", saying it "should be canceled" if corrections are unsuccessful. However, Gates has stated his support for the program. Some private analysts, such as Richard Aboulafia, of the Teal Group state that the whole F-35 program is becoming a money pit. However, on 20 January 2012, Gates' successor, Leon Panetta, lifted the F-35B's probation, stating "The STOVL variant has made — I believe and all of us believe — sufficient progress."

Former Pentagon manager Paul G. Kaminski has said that the lack of a complete test plan has added five years to the JSF program. As of February 2011, the main flaws with the aircraft are engine "screech", transonic wing roll-off and display flaws in the helmet-mounted display.

The current schedule has the delivery of basic combat capability aircraft in late 2015, followed by full capability block three software in late 2016. The $56.4 billion development project for the aircraft should be completed in 2018 when the block five configuration is expected to be delivered, several years late and considerably over budget.

Delays in the F-35 program may lead to a "fighter gap" where America and other countries will lack sufficient jet fighters to cover their requirements. Israel may seek to buy second-hand F-15s to cover its gap, while Australia may also seek to buy more American fighters from the USN to cover their own capability gap in the face of F-35 delays.

Initial Operational Capability (IOC) will be determined by software development rather than by hardware production or pilot training.

In May 2011, the Pentagon's top weapons buyer Ashton Carter said that its new $133 million unit price was not affordable.

In 2011, The Economist warned that the F-35 was in danger of slipping into a "death spiral" where increasing per aircraft costs would lead to cuts in number of aircraft ordered which would lead to further cost increases and further order cuts. Later that year, four aircraft were cut from the fifth LRIP order to pay for cost overruns.

And, in 2012, a further two aircraft were cut. Lockheed acknowledged that the slowing of purchases would increase the costs.

David Van Buren, acquisition chief for the U.S. Air Force, said that Lockheed would need to cut infrastructure to match the reduced market for their aircraft. Lockheed has said that the slowdown in American orders will free up capacity to meet the urgent short term needs of foreign partners for replacement fighters. But Air Force Secretary Michael Donley said that there was no more money available for the project and that future price increases would be matched with cuts in the number of aircraft ordered. Later that month, the Pentagon reported that costs had risen another 4.3 percent, partially as a result of production delays. In 2012, the purchase of six out of 31 aircraft was tied to performance metrics of the program.

Japan has warned that it may halt their purchase if the unit costs increase, and Canada has indicated it has not fully committed to purchasing the aircraft. The United States is projected to spend an estimated US$323 billion for development and procurement on the F-35 program, making it the most expensive defense program ever. The total lifecycle cost for the entire American fleet is estimated to be US$1.51 trillion over its 50-year life, or $618 million per plane. Testifying before a Canadian parliamentary committee in 2011, Rear Admiral Arne Røksund of Norway estimated that his country's 52 F-35 fighter jets will cost $769 million each over their operational lifetime.

Also, in 2011, a Congressional Joint Strike Fighter Caucus was formed by some of the top recipients of Lockheed Martin contributions.

James Jay Carafano of the Heritage Foundation has suggested that it would be cheaper to build additional F-35s with known defective structures and fix these later than it would be to refit legacy aircraft to remain operational until full production of F-35s with full lifespan rated structural components could be built.

The program delays have affected the program's worldwide supply chain, causing Australian Quickstep Holdings to struggle for capital, in spite of their 20-year contract with Lockheed Martin.

In order to reduce the estimated $1 trillion cost of the F-35 program over its 50-year lifetime, the USAF is considering reducing Lockheed's role in Contractor Logistics Support for the fighter. Lockheed has responded that the trillion dollar estimate relies on future costs beyond its control such as USAF reorganizations and upgrades to the aircraft that have yet to be specified.

In 2012, in order to avoid further redesign delays, the U.S. DoD accepted a reduced combat radius for the F-35A and a longer takeoff run for the F-35B. The F-35B's estimated radius has also decreased 15 percent from initial JSF goal. In a meeting in Sydney in March, the United States pledged to eight partner nations that there would be no more program delays.

On 31 May 2012, Lockheed Chief Executive Bob Stevens complained that the Defense Department's requirements for cost data were driving up the cost of the program. The same week he admitted that a strike over the company's plans to strip benefits from workers might cause a shortfall in the plan to produce 29 F-35s that year. The striking union workers raised questions about the standards of the replacement workers, even as their own work had been cited for "inattention to production quality" with a 16% rework rate. The workers went on strike to protect pensions whose costs have been the subject of negotiations with the Department of Defense over the costs of the next batch of aircraft. These same pension costs were cited by Fitch in their downgrade of the outlook for Lockheed Martin's stock price. Stevens said that while he hoped to bring down the costs of the program, the industrial base was not capable of meeting the government's expectations of affordability.

He was however able to force concessions on the workers while retaining top management's generous bonuses and pensions.

According to the latest Government Accountability Office report, the F-35's per unit cost has almost doubled, an increase of 93% over the program's 2001 baseline cost estimates. However Lockheed fears that if the tighter policies for award fees of the Obama administration are not reversed, their profits on the aircraft will be reduced by $500 million over the next five years. This was demonstrated in 2012 when the Pentagon withheld the maximum $47 million allowed for Lockheed's failure to certify its program to track project costs and schedules. The GAO has also faulted the USAF and USN for not fully planning out the full costs required to extend legacy F-16 and F-18 fleets to cover for delays in the F-35 program.

In 2006, the F-35 was downgraded from "very low observable" to "low observable", a change former RAAF flight test engineer Peter Goon likened to increasing the radar cross section from a marble to a beach ball. A Parliamentary Inquiry asked what was the re-categorization of the terminology in the United States such that the rating was changed from Very Low Observable to Low Observable.

The Department of Defence said that the change in categorization by the U.S. was due to a revision in procedures for discussing stealth platforms in a public document. The previous decision to re-categorize in the public domain has now been reversed. Publicly released material now categorizes JSF as Very Low Observable (VLO).

The spark that set off the row of criticism was US Ambassador Tom Schieffer's confirmation to the Joint Standing Committee on Foreign Affairs (JSCFADT) on 21 June 2004: With regard to the stealth technology, the airplane that Australia will get will be the stealthiest airplane that anybody outside the United States can acquire.

We have given assurances to Australia that we will give you the absolute maximum that we can with regard to that technology. Having said that, the airplane will not be exactly the same airplane as the United States will have. But it will be a stealth fighter; it will have stealth capabilities; and it will be at the highest level that anyone in the world has outside the United States.

The F-35's stealth would be similar to contemporary aircraft. A follow on article in Janes Defence Weekly, April 26, 2006, by Tom Burbage from Lockheed Martin states that the US will not export key technologies such as stealth. Before, a Jane's article gave a hint that US$1B, spent on a number of contracts, would serve to protect stealth technology and probably provide for a less stealthy export configuration of the fighter.

In response to Air Power Australia's criticisms, Australia's Air Vice Marshal Osley said that "Air Power Australia (Kopp and Goon) claim that the F35 will not be competitive in 2020 and that Air Power Australia's criticisms mainly centre around F35's aerodynamic performance and stealth capabilities." Osley continued with, "these are inconsistent with years of detailed analysis that has been undertaken by Defence, the JSF program office, Lockheed Martin, the U.S. services and the eight other partner nations.

While aircraft developments such as the Russian PAK-FA or the Chinese J20, as argued by Airpower Australia, show that threats we could potentially face are becoming increasingly sophisticated, there is nothing new regarding development of these aircraft to change Defence's assessment." He then said that he thinks that the Air Power Australia's "analysis is basically flawed through incorrect assumptions and a lack of knowledge of the classified F-35 performance information."

Andrew Krepinevich has questioned the reliance on "short range" aircraft like the F-35 or F-22 to "manage" China in a future conflict and has suggested reducing the number of F-35s ordered in favor of a longer range platform like the Next-Generation Bomber, but Michael Wynne, then United States Secretary of the Air Force rejected this plan of action in 2007. However in 2011, the Center for Strategic and Budgetary Assessments (CSBA) pointed to the restructuring of the F-35 program and the return of the bomber project as a sign of their effectiveness, while Rebecca Grant said that the restructuring was a "vote of confidence" in the F-35 and "there is no other stealthy, survivable new fighter program out there". Lockheed has also said that the F-35 is designed to launch internally carried bombs at supersonic speed and internal missiles at maximum supersonic speed.

In 2008, it was reported that RAND Corporation conducted simulated war games in which Russian Sukhoi Su-35 fighters defeated the F-35. As a result of these media reports, then Australian defence minister Joel Fitzgibbon requested a formal briefing from the Australian Department of Defence on the simulation. This briefing stated that the reports of the simulation were inaccurate and did not actually compare the F-35's flight performance against other aircraft.

The Pentagon and Lockheed Martin added that these simulations did not address air-to-air combat. A Lockheed Martin press-release points to USAF simulations regarding the F-35's air-to-air performance against potential adversaries described as "4th generation" fighters, in which it claims the F-35 is "400 percent" more effective. Major General Charles R. Davis, USAF, the F-35 program executive officer, has stated that the "F-35 enjoys a significant Combat Loss Exchange Ratio advantage over the current and future air-to-air threats, to include Sukhois". The nature of the simulations, and the terms upon which the "400 percent" figure have been derived remains unclear.

In March 2012, Tom Burbage, and Gary Liberson, of Lockheed Martin addressed an Australian Parliamentary Committee about earlier assessments. They stated "Time has moved on since 2008 and we know a lot more about this airplane now than we knew then. ... Our current assessment that we speak of is greater than 6 to 1 relative loss exchange ratio against, in 4 versus 8 engagement scenarios—4 blue F-35s versus 8 advanced red threats in the 2015 to 2020 time frame. And it is very important to note that is without the pilot in the loop and are the lowest number that we talk about, the greater than 6 to 1 is when we include the pilot in the loop simulator activities".

They said: "we actually have a fifth-gen airplane flying today. The F22 has been in many exercises and is much better than the simulations forecast. We have F35 flying today; it has not been put into that scenario yet, but we have very high quality information on the capability of the sensors and the capability of the airplane, and we have represented the airplane fairly and appropriately in these large-scale campaign models that we are using. But it is not just us—it is our air force; it is your air force; it is all the other participating nations that do this; it is our navy and our marine corps that do these exercises. It is not Lockheed in a closet gleaning up some sort of result." Although the advanced threats are classified they indicated that all the first-tier air forces in the world would not look at analysis against inferior threats."

Regarding the original plan to fit the F-35 with only two air-to-air missiles (internally), Major Richard Koch, chief of USAF Air Combat Command's advanced air dominance branch is reported to have said that "I wake up in a cold sweat at the thought of the F-35 going in with only two air-dominance weapons." However the Norwegians have been briefed on a plan to equip the F-35 with six AIM-120D missiles by 2019.

Former RAND author John Stillion has written of the F-35A's air-to-air combat performance that it "can't turn, can't climb, can't run", but Lockheed Martin test pilot Jon Beesley has countered that in an air-to-air configuration the F-35 has almost as much thrust as weight and a flight control system that allows it to be fully maneuverable even at a 50-degree angle of attack.

Andrew Hoehn, Director of RAND Project Air Force, made the following statement: "Recently, articles have appeared in the Australian press with assertions regarding a war game in which analysts from the RAND Corporation were involved. Those reports are not accurate. RAND did not present any analysis at the war game relating to the performance of the F-35 Joint Strike Fighter, nor did the game attempt detailed adjudication of air-to-air combat. Neither the game nor the assessments by RAND in support of the game undertook any comparison of the fighting qualities of particular fighter aircraft."

In an interview with the state-run Global Times, Chen Hu, editor-in-chief of World Military Affairs magazine has said that the F-35 is too costly because it attempts to provide the capabilities needed for all three American services in a common airframe. Dutch news program NOVA show interviewed U.S. defense specialist Winslow T. Wheeler and aircraft designer Pierre Sprey who called the F-35 "heavy and sluggish" as well as having a "pitifully small load for all that money", and went on to criticize the value for money of the stealth measures as well as lacking fire safety measures. His final conclusion was that any air force would be better off maintaining its fleets of F-16s and F/A-18s compared to buying into the F-35 program. Lockheed spokesman John Kent has said that the missing fire-suppression systems would have offered "very small" improvements to survivability.

In the context of selling F-35s to Israel to match the F-15s that will be sold to Saudi Arabia, a senior U.S. defense official was quoted as saying that the F-35 will be "the most stealthy, sophisticated and lethal tactical fighter in the sky," and added "Quite simply, the F-15 will be no match for the F-35." After piloting the aircraft, RAF Squadron Leader Steve Long said that, over its existing aircraft, the F-35 will give "the RAF and Navy a quantum leap in airborne capability."

Consultant to Lockheed Martin Loren B. Thompson has said that the "electronic edge F-35 enjoys over every other tactical aircraft in the world may prove to be more important in future missions than maneuverability".

In 2011, Canadian politicians raised the issue of the safety of the F-35's reliance on a single engine (as opposed to a twin-engine configuration, which provides a backup in case of an engine failure). Canada, and other operators, had previous experience with a high-accident rate with the single-engine Lockheed CF-104 Starfighter with many accidents related to engine failures. Defence Minister Peter MacKay, when asked what would happen if the F-35's single engine fails in the Far North, stated "It won't".

In November 2011, a Pentagon study team identified the following 13 areas of concern that remained to be addressed in the F-35:

The helmet-mounted display system does not work properly.

The fuel dump subsystem poses a fire hazard.

The Integrated Power Package is unreliable and difficult to service.

The F-35C's arresting hook does not work.

Classified "survivability issues", which have been speculated to be about stealth.

The wing buffet is worse than previously reported.

The airframe is unlikely to last through the required lifespan.

The flight test program has yet to explore the most challenging areas.

The software development is behind schedule.

The aircraft is in danger of going overweight or, for the F-35B, not properly balanced for VTOL operations.

There are multiple thermal management problems. The air conditioner fails to keep the pilot and controls cool enough, the roll posts on the F-35B overheat, and using the afterburner damages the aircraft.

The automated logistics information system is partially developed.

The lightning protection on the F-35 is uncertified, with areas of concern.

In December 2011 the Pentagon and Lockheed came to an agreement to assure funding and delivery for a fifth order of early F-35 aircraft of yet undefined type in spite of general national austerity measures affecting the program.

Michael Auslin of the American Enterprise Institute has questioned the capability of the F-35 to engage modern air defenses, in spite of Russia's own admission that the S-300 systems are vulnerable to the F-35.

In July 2012, the Pentagon awarded Lockheed another $450 million to fix the electronic warfare systems of the F-35.

The F-35 appears to be a smaller, slightly more conventional, single-engine sibling of the sleeker, twin-engine Lockheed Martin F-22 Raptor, and indeed drew elements from it. The exhaust duct design was inspired by the General Dynamics Model 200 design, which was proposed for a 1972 supersonic VTOL fighter requirement for the Sea Control Ship. For specialized development of the F-35B STOVL variant, Lockheed consulted with the Yakovlev Design Bureau, purchasing design data from their development of the Yakovlev Yak-141 "Freestyle". Although several experimental designs have been built and tested since the 1960s including the navy's unsuccessful Rockwell XFV-12, the F-35B is to be the first operational supersonic, STOVL stealth fighter.

The F-35 has a maximum speed of over Mach 1.6. With a maximum takeoff weight of 60,000 lb (27,000 kg), the Lightning II is considerably heavier than the lightweight fighters it replaces. In empty and maximum gross weights, it more closely resembles the single-seat, single-engine Republic F-105 Thunderchief, which was the largest single-engine fighter of the Vietnam war era. The F-35's modern engine delivers over 60 percent more thrust in an aircraft of the same weight so that in thrust to weight and wing loading it is much closer to a comparably equipped F-16.

Acquisition deputy to the assistant secretary of the air force, Lt. Gen. Mark D. "Shack" Shackelford has said that the F-35 is designed to be America's "premier surface-to-air missile killer and is uniquely equipped for this mission with cutting edge processing power, synthetic aperture radar integration techniques, and advanced target recognition."

Some improvements over current-generation fighter aircraft are:

Durable, low-maintenance stealth technology, using structural fiber mat instead of the high-maintenance coatings of legacy stealth platforms;

Integrated avionics and sensor fusion that combine information from off- and on-board sensors to increase the pilot's situational awareness and improve target identification and weapon delivery, and to relay information quickly to other command and control (C2) nodes; High speed data networking including IEEE 1394b and Fibre Channel. (Fibre Channel is also used on Boeing's Super Hornet.)

The Autonomic Logistics Global Sustainment (ALGS), Autonomic Logistics Information System (ALIS) and Computerized Maintenance Management System (CMMS) are claimed to help ensure aircraft uptime with minimal maintenance manpower. However the Pentagon has moved to open the sustainment for competitive bidding by other companies. This was after Lockheed admitted that instead of costing twenty percent less than the F-16 per flight hour, the F-35 would actually cost twelve percent more. The USMC have implemented a workaround for a cyber vulnerability in the system. Electrohydrostatic actuators run by a power-by-wire flight-control system.

Lockheed Martin claims the F-35 is intended to have close and long-range air-to-air capability second only to that of the F-22 Raptor. The company has suggested that the F-35 could also replace the USAF's F-15C/D fighters in the air superiority role and the F-15E Strike Eagle in the ground attack role, but it does not have the range or payload of either F-15 model. The F-35A does carry a similar air-to-air armament as the conceptual Boeing F-15SE Silent Eagle when both aircraft are configured for low observable operations and has over 80 percent of the larger aircraft's combat radius, under those conditions.

F-35A prototype being towed to its inauguration ceremony on 7 July 2006

Lockheed Martin has said that the F-35 has the advantage over the F-22 in basing flexibility and "advanced sensors and information fusion".

Structural composites in the F-35 are 35% of the airframe weight (up from 25% in the F-22). The majority of these are bismaleimide (BMI) and composite epoxy material. However the F-35 will be the first mass produced aircraft to include structural nanocomposites, namely carbon nanotube reinforced epoxy.

The F-35 program has learned from the corrosion problems that the F-22 had when it was first introduced in 2005. The F-35 uses a gap filler that causes less galvanic corrosion to the skin, is designed with fewer gaps in its skin that require gap filler, and has better drainage.

A United States Navy study found that the F-35 will cost 30 to 40 percent more to maintain than current jet fighters. A Pentagon study found that it may cost $1 trillion to maintain the entire fleet over its lifetime.

The relatively short 35 foot wingspan of the A and B variants is set by the F-35B's requirement to fit inside the Navy's current amphibious assault ship elevators. The F-35C's longer wing is considered to be more fuel efficient.

The F-35's main engine is the Pratt & Whitney F135. The General Electric/Rolls-Royce F136 was under development as an alternative engine until December 2011 when the manufacturers canceled work on it. The F135/F136 engines are not designed to supercruise in the F-35, but the F-35 can achieve a limited supercruise of Mach 1.2 for 150 miles.

The STOVL versions of both power plants use the Rolls-Royce LiftSystem, designed by Lockheed Martin and developed to production by Rolls-Royce. This system is more like the Russian Yak-141 and German VJ 101D/E than the preceding generation of STOVL designs, such as the Harrier Jump Jet in which all of the lifting air went through the main fan of the Rolls-Royce Pegasus engine.

The Lift System is composed of a lift fan, drive shaft, two roll posts and a "Three Bearing Swivel Module" (3BSM). The 3BSM is a thrust vectoring nozzle which allows the main engine exhaust to be deflected downward at the tail of the aircraft.

The lift fan is near the front of the aircraft and provides a counterbalancing thrust using two counter-rotating blisks.

It is powered by the engine's low-pressure (LP) turbine via a drive shaft and gearbox. Roll control during slow flight is achieved by diverting unheated engine bypass air through wing-mounted thrust nozzles called Roll Posts.

Like lift engines, the added lift fan machinery increases payload capacity during vertical flight, but is dead weight during horizontal flight. The cool exhaust of the fan also reduces the amount of hot, high-velocity air that is projected downward during vertical takeoff, which can damage runways and aircraft carrier decks.

To date, F136 funding has come at the expense of other parts of the program, reducing the number of aircraft built and increasing their costs. The F136 team has claimed that their engine has a greater temperature margin which may prove critical for VTOL operations in hot, high altitude conditions.

Pratt & Whitney is also testing higher thrust versions of the F135, partly in response to GE's claims that the F136 is capable of producing more thrust than the 43,000 lbf (190 kN) supplied by early F135s. The F135 has demonstrated a maximum thrust of over 50,000 lbf (220 kN) during testing. The F-35's Pratt & Whitney F135 is the most powerful engine ever installed in a fighter aircraft.

The F135 is the second (radar) stealthy afterburning jet engine and like the Pratt & Whitney F119 from which it was derived, has suffered from pressure pulsations in the afterburner at low altitude and high speed or "screech". In both cases this problem was fixed during development of the fighter program.

Turbine bearing health in the engine will be monitored with thermoelectric powered wireless sensors.

The F-35A includes a GAU-22/A, a four-barrel version of the GAU-12 Equalizer 25 mm cannon. The cannon is mounted internally with 182 rounds for the F-35A or in an external pod with 220 rounds for the F-35B and F-35C. The gun pod for the B and C variants will have stealth features.

The Terma A/S multi-mission pod (MMP) could be used for different equipment in the future for all three variants, such as electronic warfare equipment, reconnaissance equipment, or possibly a rearward-facing radar.

It has two internal weapons bays, and external hard points that can mount four under wing pylons and two near wingtip pylons. The two outer hard points can only carry pylons for the AIM-9X Sidewinder and AIM-132 ASRAAM short-range air-to-air missiles (AAM). The other pylons can carry the AIM-120 AMRAAM BVR AAM, Storm Shadow air-launched cruise missile, AGM-158 Joint Air to Surface Stand-off Missile (JASSM) cruise missile, and guided bombs. The external pylons can carry missiles, bombs, and fuel tanks at the expense of reduced stealth. An air-to-air load of eight AIM-120s and two AIM-9s is possible using internal and external weapons stations; a configuration of six 2,000 lb (910 kg) bombs, two AIM-120s and two AIM-9s can also be arranged.

Internally, up to two 2,000 lb (910 kg) air-to-ground bombs can be carried in A and C models (two 1,000 lb (450 kg) bombs in the B model,) along with two smaller weapons, normally expected to be air-to-air missiles. The weapon bays can carry AIM-120 AMRAAM, AIM-132 ASRAAM, the Joint Direct Attack Munition (JDAM) – up to 2,000 lb (910 kg), the Joint Stand off Weapon (JSOW), Brimstone anti-armor missiles, and Cluster Munitions (WCMD).

Lockheed Martin states that the weapons load can also be configured as all-air-to-ground or all-air-to-air, and has suggested that a Block 5 version will be able to carry three internal weapons per bay instead of two, replacing the heavy bomb with two smaller weapons such as AIM-120 AMRAAM air-to-air missiles. Upgrades include up to four GBU-39 Small Diameter Bombs (SDB) in each bay (three per bay in F-35B, or four GBU-53/B in each bay for all F-35 variants;

The MBDA Meteor air-to-air missile is currently being adapted to fit four internally in the missile spots and may be integrated into the F-35. A modified Meteor design with smaller tailfins for the F-35 was revealed in September 2010. The United Kingdom had originally planned to put up to four AIM-132 ASRAAM internally but this has been changed to carry two internal and two external ASRAAMs. The external ASRAAMs are planned to be carried on "stealthy" pylons to increase the F-35's radar cross section slightly; the missile allow attacks to slightly beyond visual range without using radar that might alert the target.

Norway and Australia are funding a program to adapt the Naval Strike Missile (NSM) to fit the internal bays of the F-35. This will be a multi-role version, named the Joint Strike Missile (JSM), and will be the only cruise missile to fit the internal bays. Studies have shown that the F-35 would be able to carry two of these internally, while four additional missiles could be carried externally. The missile has an expected range in excess of 150 nmi (278 km).

Solid-state lasers were being developed as optional weapons for the F-35 as of 2002. The F-35 is expected to take on the Wild Weasel mission, but there are no planned anti-radiation missiles for internal stealthy carriage. The B61 nuclear bomb was scheduled for deployment in 2017, but delays in the F-35 program may delay this. It is now expected to be deployable in the early 2020s.

Weapons bay on a mock-up of the F-35

The F-35 has been designed to have a low radar cross section primarily due to the shape of the aircraft and the use of stealthy materials used in construction, including fiber-mat. Unlike the previous generation of fighters, the F-35 was designed with a shape for low-observable characteristics.

The Teen Series of fighters (F-15, F-16, F/A-18) were notable for always carrying large external fuel tanks, but in order to avoid negating its stealth characteristics the F-35 must fly most missions without external fuel tanks. Unlike the F-16 and F/A-18, the F-35 lacks leading edge extensions and instead uses stealth-friendly chines for vortex lift in the same fashion as the SR-71 Blackbird.

The small bumps just forward of the engine air intakes form part of the diverter less supersonic inlet (DSI) which is a simpler, lighter means to ensure high-quality airflow to the engine over a wide range of conditions. These inlets also crucially improve the aircraft's low-observable characteristics.

In spite of being smaller than the F-22, the F-35 has a larger radar cross section. It is said to be roughly equal to a metal golf ball rather than the F-22's metal marble. The F-22 was designed to be difficult to detect by all types of radars and from all directions. The F-35 on the other hand manifests its lowest radar signature from the frontal aspect because of compromises in design. Its surfaces are shaped to best defeat radars operating in the X and upper S band, which are typically found in fighters, surface-to-air missiles and their tracking radars, although the aircraft would be easier to detect using other radar frequencies. Because the shape of the aircraft is so important to its radar cross section, special care must be taken to maintain the "outer mold line" during production. Ground crews require Repair Verification Radar (RVR) test sets in order to verify the RCS of the aircraft after performing repairs, which was not a concern for previous generations of non-stealth fighters.

Low observable aircraft must consider different types of detection and so the F-35 is not only radar stealthy, but it also has infrared and visual signature reduction incorporated.

In late 2008 the air force revealed that the F-35 would be about twice as loud at takeoff as the McDonnell Douglas F-15 Eagle and up to four times as loud during landing. As a result, residents near Luke Air Force Base, Arizona and Eglin Air Force Base, Florida, possible home bases for the jet, requested that the air force conduct environmental impact studies concerning the F-35's noise levels. The city of Valparaiso, Florida, adjacent to Eglin AFB, threatened in February 2009 to sue over the impending arrival of the F-35s, but this lawsuit was settled in March 2010. Moreover, it was reported in March 2009 that testing by Lockheed Martin and the Royal Australian Air Force revealed that the F-35 was not as loud as first reported, being "only about as noisy as an F-16 fitted with a Pratt & Whitney F100-PW-200 engine" and "quieter than the Lockheed Martin F-22 Raptor and the Boeing F/A-18E/F Super Hornet."

However according to an acoustics study done by Lockheed Martin and the U.S. Air Force, the noise levels of the F-35 are found to be comparable to the F-22 Raptor and F/A-18E/F Super Hornet. And a USAF environmental impact study found that replacing the F-16s with F-35s at Tucson International Airport would subject more than 21 times as many residents to extreme noise levels.

The F-35 features a full-panel-width "panoramic cockpit display" (PCD) glass cockpit, with dimensions of 20 by 8 inches (50 by 20 centimeters). A cockpit speech-recognition system (Direct Voice Input) provided by Adacel is planned to improve the pilot's ability to operate the aircraft over the current-generation interface.

The F-35 will be the first U.S. operational fixed-wing aircraft to use this system, although similar systems have been used in AV-8B and trialled in previous U.S. jets, particularly the F-16 VISTA.

A helmet-mounted display system (HMDS) will be fitted to all models of the F-35. A helmet-mounted cueing system is already in service with the F-15s, F-16s and F/A-18s. While some fighters have offered HMDS along with a head up display (HUD), this will be the first time in several decades that a front line tactical jet fighter has been designed without a HUD. The F-35 is equipped with a right-hand HOTAS side stick controller.

The Martin-Baker US16E ejection seat is used in all F-35 variants. The US16E seat design balances major performance requirements, including safe-terrain-clearance limits, pilot-load limits, and pilot size. It uses a twin-catapult system that is housed in side rails. The F-35 uses a derivative version of the oxygen system that has been implicated in hypoxia incidents on board the F-22. But the F-35 does not fly as high or as fast as the F-22; its flight profile is similar to other fighters that use such systems routinely.

The F-35's sensor and communications suite is intended to facilitate situational awareness, command-and-control and network-centric warfare capability. The main sensor on board the F-35 is its AN/APG-81 AESA-radar, designed by Northrop Grumman Electronic Systems. It is augmented by the Electro-Optical Targeting System (EOTS) mounted under the nose of the aircraft, designed by Lockheed Martin. This gives the same capabilities as the Lockheed Martin Sniper XR while avoiding making the aircraft more easily detectable.

Six additional passive infrared sensors are distributed over the aircraft as part of Northrop Grumman's AN/AAQ-37 distributed aperture system (DAS), which acts as a missile warning system, reports missile launch locations, detects and tracks approaching aircraft spherically around the F-35, and replaces traditional night vision goggles for night operations and navigation. All DAS functions are performed simultaneously, in every direction, at all times. The F-35's Electronic Warfare systems are designed by BAE Systems and include Northrop Grumman components. Some functions such as the Electro-Optical Targeting System and the Electronic Warfare system are not usually found integrated on fighters.

The AN/ASQ-239 (Barracuda) system is an improved version of the AN/ALR-94 EW suite on the F-22. The AN/ASQ-239 provides sensor fusion of RF and IR tracking functions, basic radar warning, multispectral countermeasures for self-defense against threat missiles, situational awareness and electronic surveillance. It uses 10 RF antennae over the leading and trailing edges of the wing leading and trailing edges of the horizontal tail.

The communications, navigation and identification (CNI) suite is designed by Northrop Grumman and includes the Multifunction Advanced Data Link (MADL). The F-35 will be the first jet fighter that has sensor fusion that combines both radio frequency and IR tracking for continuous target detection and identification in all directions which is shared via MADL to other platforms without compromising low observability. However the F-35 also includes the non-stealthy Link 16 for communications with legacy systems for missions including Close air support.

The F-35 has been designed with synergy between sensors as a specific requirement, with the "senses" of the aircraft expected to provide a more cohesive picture of the reality around it, and be available in principle for use in any possible way, and any possible combination with one another.

All of the sensors feed directly into the main processors to support the entire mission of the aircraft. For example the AN/APG-81 functions not just as a multi-mode radar, but also as part of the aircraft's electronic warfare system. Northrop Grumman is offering the APG-81 as an upgrade for legacy aircraft, but because the "back end processing" on the F-35 is done in software on the main processors, the upgrade version requires their Scalable Agile Beam Radar electronics to operate on other aircraft.

Unlike previous aircraft, such as the F-22, all software for the F-35 is written in C++ for faster code development. The Integrity DO-178B real-time operating system (RTOS) from Green Hills Software runs on COTS Freescale PowerPC processors. The final Block 3 software for the F-35 is planned to have 8.6 million lines of software code.

The scale of the program has led to a software crisis as officials continue to discover that additional software needs to be written. General Norton Schwartz has said that the software is the biggest factor that might delay the USAF's initial operational capability which is now scheduled for April 2016. Michael Gilmore, Director of Operational Test & Evaluation, has written that, "the F-35 mission systems software development and test is tending towards familiar historical patterns of extended development, discovery in flight test, and deferrals to later increments."

The F-35's electronic warfare systems are intended to detect hostile aircraft first, which can then be scanned with the electro-optical system and action taken to engage or evade the opponent before the F-35 is detected. The CATbird avionics testbed for the F-35 program has proved capable of detecting and jamming F-22 radars.

The F-35 was previously considered a platform for the Next Generation Jammer, but attention has shifted to the use of unmanned platforms.

The F-35 does not need to be physically pointing at its target for weapons to be successful. This is possible because of sensors that can track and target a nearby aircraft from any orientation, provide the information to the pilot through his helmet (and therefore visible no matter which way they are looking), and provide the seeker-head of a missile with sufficient information. Recent missile types provide a much greater ability to pursue a target regardless of the launch orientation, called "High Off-Boresight" capability, although the speed and direction in which the munition is launched affect the effective range of the weapon. Sensors use combined radio frequency and infra red (SAIRST) to continually track nearby aircraft while the pilot's helmet-mounted display system (HMDS) displays and selects targets. The helmet system replaces the display suite-mounted head-up display used in earlier fighters.

The F-35's systems provide the edge in the "observe, orient, decide, and act" OODA loop; stealth and advanced sensors aid in observation (while being difficult to observe), automated target tracking helps in orientation, sensor fusion simplifies decision making, and the aircraft's controls allow the pilot to keep their focus on the targets, rather than the controls of their aircraft.

The problems with the current Vision Systems International helmet-mounted display led Lockheed Martin to issue a draft specification for proposals for an alternative on 1 March 2011. The alternative system will be based on Anvis-9 night vision goggles. It will be supplied by BAE systems. The BAE system does not yet include all the features of the VSI helmet and if successful will have the remaining features incorporated. Use of the BAE system would also require a cockpit redesign.

In 2011, Lockheed granted VSI a contract to fix the vibration, jitter, night-vision and sensor display problems in their helmet-mounted display. The improved displays are expected to be delivered in third quarter of 2013. One of the potential improvements is to replace Intevac's ISIE-10 day/night camera located in the helmet with their ISIE-11 model which will improve the resolution from 1280x1024 to 1600x1200 pixels. By October 2012, Lockheed said it had made progress in resolving technical issues of the helmet-mounted display, citing positive reports from night flying tests. Night vision performance was the "only real question" left on the helmet, and there was question as to whether the helmet system would allow pilots to see well enough at night to carry out precision tasks.

The program's maintenance concept is for any F-35 to be maintained in any F-35 maintenance facility and that all F-35 parts in all bases will be globally tracked and shared as needed. The commonality between the different variants has allowed the USMC to create their first aircraft maintenance Field Training Detachment to directly apply the lessons of the USAF to their own F-35 maintenance operations.

The aircraft has been designed for ease of maintenance, with 95% of all field replaceable parts "one deep" where nothing else has to be removed to get to the part in question. For instance the ejection seat can be replaced without removing the canopy, the aircraft uses low-maintenance electro-hydrostatic actuators instead of hydraulic systems and an all-composite skin without the fragile coatings found on earlier stealth aircraft.

The F-35 has received good reviews from pilots and maintainers, suggesting it is performing better than its predecessors did at a similar stage of development. The stealth type has proved relatively stable from a maintenance standpoint. Occasionally, there some issues with it on the ground, however, it is usually easily fixed by shutting the aircraft down and restarting it. Part of the improvement is attributed to better maintenance training, as F-35 maintainers have received far more extensive instruction at this early stage of the program than on the F-22 Raptor. Furthermore, the F-35's stealth coatings are much easier to work with than those used on the Raptor. Cure times for coating repairs are lower and many of the fasteners and access panels are not coated, further reducing the workload for maintenance crews. Some of the F-35's radar-absorbent materials are baked into the jet's composite skin, which means its stealthy signature is not easily degraded. However, it is still harder to maintain (due to its stealth) than fourth-generation aircraft.

The first F-35A (designated AA-1) was rolled out in Fort Worth, Texas on 19 February 2006. The aircraft underwent extensive ground testing at Naval Air Station Joint Reserve Base Fort Worth in late 2006. In September 2006 the first engine run of the F135 afterburner turbofan in an airframe and tests were completed; the first time that the F-35 was completely functional on its own power systems. On 15 December 2006, the F-35A completed its maiden flight. A modified Boeing 737–300, the Lockheed CATBird is used as an avionics test bed inside of which are racks holding all of F-35's avionics, as well as a complete F-35 cockpit.

On 31 January 2008 at Fort Worth, Texas, Lieutenant Colonel James "Flipper" Kromberg of the U.S. Air Force became the first military service pilot to evaluate the F-35, taking the aircraft through a series of maneuvers on its 26th flight. F-35 AA-1, on its 34th test flight, began aerial refueling testing in March 2008. Another milestone was reached on 13 November 2008, when the AA-1 flew at supersonic speeds for the first time, reaching Mach 1.05 at 30,000 ft (9,144 m) making four transitions through the sound barrier, for a total of eight minutes of supersonic flight.

The first F-35B (designated BF-1) made its maiden flight on 11 June 2008. The flight, which featured a conventional takeoff, was piloted by BAE Systems' test pilot Graham Tomlinson. The BF-1 is the second of 19 System Development and Demonstration (SDD) F-35s, and the first to use new weight-optimized design features that will apply to all future F-35s. Testing of the STOVL propulsion system in flight began on 7 January 2010.

The STOVL system was used for 14 minutes of the 48-minute test flight while the aircraft slowed from 210 knots (390 km/h) to 180 knots (330 km/h). The F-35B's first hover (full stop in mid-air) happened on 17 March 2010, followed by a STOVL landing, and on 18 March 2010 the first vertical landing was performed. During a test flight on 10 June 2010, the F-35B became the second STOVL aircraft to achieve supersonic speeds, the first being its ancestor, the X-35B, which achieved the same feat on 20 July 2001. In January 2011, Lockheed Martin reported it had solved a problem with the aluminum bulkhead used only on the F-35B which had cracked during ground testing.

Although many of the initial flight test targets have been accomplished, the F-35 testing program completed "just under 100 sorties and about as many hours in 2.5 years" by June 2009 and was falling significantly behind schedule. A 2008 Pentagon Joint Estimate Team (JET I) estimated that the program was two years behind the latest public schedule, and a 2009 Joint Estimate Team (JET II) revised that estimate to predict a 30-month delay. Due to those delays in the testing program, production numbers will be reduced by 122 aircraft through 2015 in order to provide additional funds for development.

Those additional funds will add $2.8 billion to the development funds and internal memos suggest that the official timeline will be extended by 13 months (not the 30 months the JET II team predicted the slip would be).

The success of the Joint Estimate Team has led Ashton Carter to call for more such teams for other poorly performing Pentagon projects.

In June 2009, the F-35s APG-81 active electronically scanned array radar was integrated in the Northern Edge 2009 large-scale military exercise when it was mounted on the front of a Northrop Grumman test aircraft. The APG-81 test event represented a major milestone in electronic protection testing in an operationally representative environment, accomplished years ahead of normal developmental timelines and providing a significant risk reduction opportunity for the aircraft's key sensor. The test events "validated years of laboratory testing versus a wide array of threat systems, showcasing the extremely robust electronic warfare capabilities of the world's most advanced fighter fire control radar."

Nearly 30 percent of all the test flights have required more than routine maintenance to get the aircraft flyable again. As of March 2010, the F-35 program had used a million more man-hours than predicted and flight testing is expected to result in further design changes. The United States Navy has projected that lifecycle costs over a fleet life of 65 years for all of the American F-35s will be $442 billion higher than the U.S. Air Force has projected. The delay in the F-35 program is expected to lead to a shortfall of around 100 jet fighters in the Navy/Marines team. Given careful management, service life extension of the Marines' legacy F/A-18s, and more burdens placed on Navy fighters, it may be possible to reduce this shortfall.

The F-35C carrier variant's maiden flight took place on 7 June 2010, also at NAS Fort Worth JRB. The 57-minute flight was executed by Lockheed test pilot Jeff "Slim" Knowles, who was the chief test pilot for the F-117 program. A total of 11 U.S. Air Force F-35s arrived in Fiscal Year 2011.

On 9 March 2011 all F-35s were grounded after a dual generator failure and oil leak in flight. This was the first significant flight failure since 2007. Seven of the 10 test aircraft were cleared to fly four days later. These aircraft have an older model of generators, unlike the kind that failed in flight. The problem was found to be the result of faulty maintenance rather than a design or construction issue.

In June 2011, the F-35's sensors (radar and DAS) were tested in an operational exercise named Northern Edge 2011. Navy Cmdr. Erik Etz, the deputy mission systems integrated product team lead from the F-35 JSF Program Office, said the rigorous testing of both sensors during Northern Edge 2011 served as a significant risk-reduction step for the F-35 JSF program. "By putting these systems in this operationally rigorous environment, we have demonstrated key war fighting capabilities well in advance of scheduled operational testing," Commander Etz added.

From 3–18 August 2011, the F-35 fleet was grounded. The precautionary grounding was in effect while the Joint Program Office investigated the cause of an electrical system failure on test aircraft, F-35A AF-4 during ground tests. The Honeywell-built integrated power package (IPP) failed during a standard engine test following a maintenance check on 2 August at Edwards Air Force Base. The IPP is relatively unheralded, but plays a major role in the F-35's performance. It combines the functions of an environment control unit, engine starter and back-up power generator into a single system. The IPP has been a target for concern during the development phase, even though it has never failed before in more than 1,500 flight hours.

The JPO was assessing the impact of the grounding on the schedule for the system development and demonstration phase, but the latest version of the schedule included margin for unexpected problems.

On 10 August 2011, ground operations for the F-35 Program were reinstituted while the investigation was ongoing; government and contractor engineering teams, after reviewing initial data, determined that the fleet could safely resume Development Test ground operations. Preliminary root cause inquiries indicate that a control valve did not function properly, which in turn led to the IPP failure. Monitoring of this valve is the mitigating action required to allow ground operations. The F-35 team is revising ground monitoring procedures to ensure testing involving the IPP takes place safely. While initiating DT ground operations is a "major step" for the F-35 fleet returning to flight, further reviews are required prior to lifting the suspension of flight operations for the 20 F-35s currently in flying status. On 18 August 2011, the flight ban was lifted for 18 of the 20 fighters. Two aircraft based at Eglin Air Force Base in Florida will remain grounded because they lack the monitoring systems used in developmental test aircraft that can detect any problems in flight.

On 25 October 2011 the F-35A reached its designed top speed of Mach 1.6 for the first time. Further testing demonstrated a speed of Mach 1.61 and 9.9g. During testing in 2011, all eight landing tests of the F-35C tail hook failed to catch the arresting wire; the hook design is being modified to address the problem.

On 8 August 2012, an F-35B dropped an inert 1,000-pound bomb over the Atlantic test range in the program's first airborne ordnance separation test. The F-35A dropped its first bomb on 16 October 2012.

On 15 August 2012, an F-35B completed a series of engine air start tests, which involve shutting down and restarting the turbofan while in flight. The aircraft, designated BF-2, successfully completed a series of 27 air starts at various altitudes and using various methods. The F-35A variant completed its air start tests previously.

On 21 August 2012, J. Michael Gilmore wrote that he would not approve the Test and Evaluation Master Plan until his concerns about electronic warfare testing, budget and concurrency were addressed.

On 7 September 2012, the Pentagon failed to approve a comprehensive operational testing plan for the F-35. Instead on 10 September 2012, the USAF started an operational utility evaluation (OUE) of the F-35A variant. This was scheduled to last 65 training days. The purpose of the OUE was to evaluate the entire air vehicle system, including the logistical support and maintenance of it, the maintenance training, the pilot training, and the pilot execution. There were four pilots as part of the OUE. Evaluators will have 90 days to submit their report to Air Education and Training Command (AETC) chief, which should happen early in 2013. He will then either accept or decline any recommendations made in the report and is subsequently expected to formally authorize the start of training at the sea-side base. The U.S. Navy and Marines do not yet have an OUE scheduled for the F-35B and C versions. By 1 October, the OUE was reported as "proceeding smoothly." The pilots went through academics and started their normal operations simulators. Actual flying began on October 26. By 9 November, 20 out 24 planned flight sorties had been completed. The OUE was completed on November 14 with the 24th flight. The four pilots completed six flights each, along with simulator rides and an academics phase. A report on the readiness of the F-35 and the training syllabus are now being compiled. If the AETC chief finds its performance was satisfactory, normal training operations at sea-side bases will formally start.

On 19 September 2012, the F-35A completed 8,000 hours of airframe durability testing, which is equivalent to one lifetime for the aircraft. The durability testing proved the airframe was able to handle a variety of flying conditions it would experience when in service. The testing was completed ahead of schedule. On 2 October 2012, an F-35B conducted the first air-to-air refueling of an F-35 aircraft. Four aircraft, two at a time, on two separate sorties, took on fuel in midair. Previous aerial refueling operations with the F-35 had been conducted with test aircraft.

On 19 October 2012, an F-35A completed the first aerial weapons release of an AIM-120 AMRAAM from the aircraft.

On 16 November 2012, the U.S. Marines received the first F-35B at MCAS Yuma, and the VMFA(AW)-121 unit is to be redesigned from a Boeing F/A-18 Hornet unit to an F-35B squadron. On 28 November 2012, an F-35C test aircraft ejected an inert 2,000 lb GBU-31 JDAM and 500 lb GBU-12 Paveway from its internal weapons bay in the first ground weapons ejections for the F-35C. It also ejected an AIM-120 AMRAAM, with a total of eleven weapon releases.

On 3 December 2012, an F-35B released an inert 500 pound GBU-12 Paveway over international waters in the Atlantic Ocean. The aircraft was traveling at 485 kn (558 mph) at an altitude of 5,000 ft. This was the second weapons release for the F-35B, and the fourth for all Lightning II aircraft.

F-35B short-takeoff from USS Wasp (LHD-1) during its first sea trials, October 2011.

Lockheed P-3 Orion

The Lockheed P-3 Orion is a four-engine turboprop anti-submarine and maritime surveillance aircraft developed for the United States Navy and introduced in the 1960s. Lockheed based it on the L-188 Electra commercial airliner. The aircraft is easily recognizable by its distinctive tail stinger or "MAD Boom", used for the magnetic detection of submarines.

Over the years, the aircraft has seen numerous design advancements, most notably to its electronics packages. The P-3 Orion is still in use by numerous navies and air forces around the world, primarily for maritime patrol, reconnaissance, anti-surface warfare and anti-submarine warfare. A total of 734 P-3s have been built, and by 2012, it will join the handful of military aircraft such as the Boeing B-52 Stratofortress which have served 50 years of continuous use with its original primary customer, in this case, the United States Navy. The U.S. Navy's remaining P-3C aircraft will eventually be replaced by the Boeing P-8A Poseidon.

A P-3C Orion of Patrol Squadron 22 (VP-22) flies over Japan, 1 December 1991.

In August 1957, the U.S. Navy called for replacement proposals for the aging twin piston engined Lockheed P2V Neptune (later redesignated P-2) and Martin P5M Marlin (later redesignated P-5) with a more advanced aircraft to conduct maritime patrol and antisubmarine warfare. Modifying an existing aircraft was expected to save on cost and allow rapid introduction into the fleet. Lockheed suggested a military version of their L-188 Electra, which was still in development and had yet to fly. In April 1958 Lockheed won the competition and was awarded an initial research and development contract in May.

The prototype YP3V-1/YP-3A, Bureau Number (BuNo) 148276 was modified from the third Electra airframe c/n 1003. The first flight of the aircraft's aerodynamic prototype, originally designated YP3V-1, was on 19 August 1958. While based on the same design philosophy as the Lockheed L-188 Electra, the aircraft was structurally different. The aircraft had 7 feet (2.1 m) less fuselage forward of the wings with an opening bomb bay, and a more pointed nose radome, distinctive tail "stinger", wing hard points, and other internal, external, and airframe production technique enhancements. The Orion has four Allison T56 turboprops which give it a top speed of 411 knots (761 km/h) comparable to the fastest propeller fighters, or even slow high-bypass turbofan jets such as the A-10 Thunderbolt II or the Lockheed S-3 Viking. Similar patrol aircraft include the Soviet Ilyushin Il-38 and the French Breguet Atlantique, while Britain adapted the jet-powered de Havilland Comet as the Hawker Siddeley Nimrod.

The first production version, designated P3V-1, was launched on 15 April 1961. Initial squadron deliveries to Patrol Squadron Eight (VP-8) and Patrol Squadron Forty Four (VP-44) at Naval Air Station Patuxent River, Maryland began in August 1962. On 18 September 1962, the U.S. military transitioned to a unified designation system for all services, with the aircraft being renamed the P-3 Orion. Paint schemes have changed from early 1960s gloss blue and white, to mid-1960s gloss white and gray, to mid-1990s flat finish low visibility gray with fewer and smaller markings. In the early 2000s, the scheme changed to a gloss gray finish with the original full-size color markings. Large size Bureau Numbers on the vertical stabilizer and squadron designations on the fuselage remained omitted.

In 1963, the U.S. Navy Bureau of Weapons (BuWeps) contracted Univac Defense Systems Division of Sperry-Rand to engineer, build and test a digital computer (then in its infancy) to interface with the many sensors and newly-developing display units of the P-3 Orion. Project A-NEW was the engineering system which, after several early trials, produced the engineering prototype, the CP-823/U, Univac 1830, Serial A-1, A-NEW MOD3 Computing System. The CP-823/U was delivered to the Naval Air Development Center (NADC) Warminster at Johnsville, Pennsylvania in 1965, and directly led to the production computers later equipped on the P-3C Orion.

Three civilian Electras were lost in fatal accidents between February 1959 and March 1960. Following the third crash the FAA restricted the maximum speed of Electras until the cause could be determined. After an extensive investigation, two of the crashes (in September 1959 and March 1960) were found to be caused by insufficiently strong engine mounts, unable to dampen a whirling motion that could affect the outboard engines. When the oscillation was transmitted to the wings, a severe vertical vibration escalated until the wings were torn from the aircraft. The company implemented an expensive modification program, labelled the Lockheed Electra Achievement Program or LEAP, in which the engine mounts and wing structures supporting the mounts were strengthened, and some wing skins replaced with thicker material. Each of the survivors of the 145 Electras built to that time was modified at Lockheed's expense at the factory, the modifications taking 20 days for each aircraft. The changes were incorporated in subsequent aircraft as they were built.

Sales of airliners were limited as the technical fix did not completely erase the "jinxed" reputation, turboprop-powered aircraft were soon replaced by faster jets. In a military role where fuel efficiency was more valued than speed, the Orion remained in service over 50 years after its 1962 introduction.

Although surpassed in production longevity by the Lockheed C-130 Hercules, 734 P-3s were produced up until 1990. Lockheed Martin opened a new P-3 wing production line in 2008 as part of its Service Life Extension Program (ASLEP) for delivery in 2010. A complete ASLEP replaces the outer wings, center wing lower section and horizontal stabilizers with new-build parts.

In the 1990s, during a U.S. Navy attempt to identify a successor aircraft to the P-3, the improved P-7 was selected over a navalized variant of the twin turbofan-powered Boeing 757, but this program was subsequently cancelled. In a second program to procure a successor, the advanced Lockheed-Martin Orion 21, another P-3 derived aircraft, lost out to the Boeing P-8 Poseidon, a Boeing 737 variant, which is due to enter service in 2013.

The P-3 has an internal bomb bay under the front fuselage which can house conventional Mark 50 torpedoes or Mark 46 torpedoes and/or special (nuclear) weapons. Additional under wing stations, or pylons, can carry other armament configurations including the AGM-84 Harpoon, AGM-84E SLAM, AGM-84H/K SLAM-ER, the AGM-65 Maverick, 127 millimeters (5.0 in) Zuni rockets, and various other sea mines, missiles, and gravity bombs. The aircraft also had the capability to carry the AGM-12 Bull pup guided missile until that weapon was withdrawn from U.S./NATO/Allied service.

The P-3 is equipped with a magnetic anomaly detector (MAD) in the extended tail. This instrument is able to detect the magnetic anomaly of a submarine in the Earth's magnetic field. The limited range of this instrument requires the aircraft to be near the submarine at low altitude. Because of this, it is primarily is used for pinpointing the location of a submarine immediately prior to a torpedo or depth bomb attack. Due to the sensitivity of the detector, electro-magnetic noise can interfere with it, so the detector is placed in P-3's fiberglass tail stinger (MAD boom), far from other electronics and ferrous metals on the aircraft.

Developed during the Cold War, the P-3's primary mission was to track Soviet Navy ballistic missile and fast attack submarines and to eliminate same in the event of full scale war. At its height, the U.S. Navy's P-3 community consisted of twenty-four active duty "Fleet" patrol squadrons home based at air stations in the states of Florida and Hawaii as well as bases which formerly had P-3 operations in Maryland, Maine, and California. There were also thirteen Naval Reserve patrol squadrons identical to their active duty "Fleet" counterparts, said Reserve "Fleet" squadrons being based in Florida, Pennsylvania, Maryland, Michigan, Massachusetts (later relocated to Maine), Illinois, Tennessee, Louisiana, California and Washington. Two Fleet Replacement Squadrons (FRS), also called "RAG" squadrons (from the historic "Replacement Air Group" nomenclature) were located in California and Florida. The since-deactivated squadron in California provided P-3 training for the Pacific Fleet, while the squadron in Florida performed the task for the Atlantic Fleet). These squadrons were also augmented by a test and evaluation squadron in Maryland, two additional test and evaluation units that were part of an air development center in Pennsylvania and a test center in California, an oceanographic development squadron in Maryland, and two active duty "special projects" units in Texas and Hawaii, the latter being slightly smaller than a typical squadron.

Reconnaissance missions in international waters led to occasions where Soviet fighters would "bump" a U.S. Navy P-3 or other P-3 operators such as the Royal Norwegian Air Force. On 1 April 2001, a midair collision between a United States Navy EP-3E ARIES II signals surveillance aircraft and a People's Liberation Army Navy (PLAN) J-8II jet fighter-interceptor resulted in an international dispute between the U.S. and the People's Republic of China (PRC) called the Hainan Island incident.

More than 40 combatant and noncombatant P-3 variants have demonstrated the rugged reliability displayed by the platform flying 12-hour plus missions 200 ft (61 m) over salt water while maintaining an excellent safety record. Versions have been developed for the National Oceanic and Atmospheric Administration (NOAA) for research and hurricane hunting/hurricane wall busting, for the U.S. Customs Service (now U.S. Customs and Border Protection) for drug interdiction and aerial surveillance mission with a rotodome adapted from the Grumman E-2 Hawkeye or an AN/APG-66 radar adapted from the General Dynamics F-16 Fighting Falcon, and for NASA for research and development.

The U.S. Navy remains the largest P-3 operator, currently distributed between a single fleet replacement (i.e., "training) patrol squadron in Florida, 12 active duty patrol squadrons distributed between bases in Florida, Washington and Hawaii, two Navy Reserve patrol squadrons in Florida and Washington, one active duty special projects patrol squadron (VPU-1) in Hawaii, and two active duty test and evaluation squadrons. One additional active duty fleet reconnaissance squadron (VQ-1) operates the EP-3 Aries signals intelligence (SIGINT) variant in Washington.

On 2 August 1990, Iraq invaded Kuwait and was poised to strike Saudi Arabia. Within 48 hours of the initial invasion, U.S. Navy P-3C aircraft were the first American forces to arrive in the area. One was a modified platform with a prototype system known as "Outlaw Hunter." Undergoing trials in the Pacific after being developed by the Navy's Space & Naval Warfare Systems Command (SPAWAR), "Outlaw Hunter" was testing a specialized over-the-horizon targeting (OTH-T) system package when it responded. Within hours of the start of the coalition air campaign, "Outlaw Hunter" detected a large number of Iraqi patrol boats and naval vessels attempting to move from Basra and Umm Qasr to Iranian waters. "Outlaw Hunter" vectored in strike elements which attacked the flotilla near Bubiyan Island destroying 11 vessels and damaging scores more. During Desert Shield, a P-3 using infrared imaging detected a ship with Iraqi markings beneath freshly-painted bogus Egyptian markings trying to avoid detection. Several days before the 7 January 1991 commencement of Operation Desert Storm, a P-3C equipped with an APS-137 Inverse Synthetic Aperture Radar (ISAR) conducted coastal surveillance along Iraq and Kuwait to provide pre-strike reconnaissance on enemy military installations. A total of 55 of the 108 Iraqi vessels destroyed during the conflict were targeted by P-3C aircraft.

The P-3 Orion's mission expanded in the late 1990s and early 2000s to include battle space surveillance both at sea and over land. The long range and long loiter time of the P-3 Orion have proved to be an invaluable asset during Operation Iraqi Freedom and Operation Enduring Freedom. It can instantaneously provide information about the battle space it can see to ground troops, particularly the U.S. Marines.

Although the P-3 is a Maritime Patrol Aircraft, armament and sensor upgrades in the Anti-surface Warfare Improvement Program (AIP) have made it suitable for sustained combat air support over land. Since the start of the current war in Afghanistan, U.S. Navy P-3 aircraft have been operating from Kandahar in that role. Royal Australian Air Force P-3 aircraft also operated there early in the war. As of February 2010, the Australian P-3 aircraft have been operating in the area for a continuous seven years.

Recently the United States Geological Survey used the Orion to survey parts of southern and eastern Afghanistan for lithium, copper, and other mineral deposits.

Lockheed S-3 Viking

The Lockheed S-3 Viking is a four-seat twin-engine jet aircraft that was used by the U.S. Navy to identify and track enemy submarines.

In the late 1990s, the S-3B's mission focus shifted to surface warfare and aerial refueling.

The Viking also provided electronic warfare and surface surveillance capabilities to the carrier battle group. A carrier-based, subsonic, all-weather, multi-mission aircraft with long range, it carried automated weapon systems, and was capable of extended missions with in-flight refueling. Because of the engines' low-pitched sound, it was nicknamed the "Hoover" after the vacuum cleaner brand.

The S-3 was retired from front-line fleet service aboard aircraft carriers by the US Navy in January 2009, with its missions being assumed by other platforms such as the P-3C Orion, SH-60 Seahawk, and F/A-18E/F Super Hornet. Several examples continue to be flown by Air Test and Evaluation Squadron THREE ZERO (VX-30) at Naval Base Ventura County / NAS Point Mugu, California for range clearance and surveillance operations on the NAVAIR Point Mugu Range, and a single example is operated by the National Aeronautics and Space Administration (NASA) at the NASA Glenn Research Center.

An S-3B Viking launches from a catapult aboard USS Abraham Lincoln.

In the mid-1960s, the U.S. Navy developed the VSX (Heavier-than-air, Anti-submarine, Experimental) requirement for a replacement for the piston-engined Grumman S-2 Tracker as an anti-submarine aircraft to fly off the Navy's aircraft carriers. In August 1968, a team led by Lockheed and a Convair/Grumman team were asked to further develop their proposals to meet this requirement. Lockheed recognised that it had little recent experience in designing carrier based aircraft, so Ling-Temco-Vought (LTV) was brought into the team, being responsible for the folding wings and tail, the engine nacelles, and the landing gear, which was derived from A-7 Corsair II (nose) and F-8 Crusader (main). Sperry Univac Federal Systems was assigned the task of developing the aircraft's onboard computers which integrated input from sensors and sonoboys.

On 4 August 1969, Lockheed's design was selected as the winner of the contest, and eight prototypes, designated YS-3A were ordered. The first prototype flew on 21 January 1972 and the S-3 entered service in 1974. During the production run from 1974 to 1978, a total of 186 S-3As were built. The majority of the surviving S-3As were later upgraded to the S-3B variant, with sixteen aircraft converted into ES-3A Shadow electronic intelligence (ELINT) collection aircraft.

The S-3 is a conventional monoplane with a high-mounted cantilever wing, swept at an angle of 15°. The two GE TF-34 high-bypass turbofan engines mounted in nacelles under the wings provide excellent fuel efficiency, giving the Viking the required long range and endurance, while maintaining docile engine-out characteristics.

The aircraft can seat four crew members, three officers and one enlisted aircrewman, with the pilot and the copilot/tactical coordinator (COTAC) in the front of the cockpit and the tactical coordinator (TACCO) and sensor operator (SENSO) in the back. Entry is by an entry door / ladder which folds out of the side of the fuselage. When the aircraft's anti-submarine warfare (ASW) role ended in the late 1990s, the enlisted SENSOs were removed from the crew. In the tanking crew configuration, the S-3B typically flew with only a crew of two (pilot and COTAC).

The wing is fitted with leading edge and Fowler flaps. Spoilers are fitted to both the upper and the lower surfaces of the wings. All control surfaces are actuated by dual hydraulically boosted irreversible systems. In the event of dual hydraulic failures, an Emergency Flight Control System (EFCS) permits manual control with greatly increased stick forces and reduced control authority.

Unlike many tactical jets which required ground service equipment, the S-3 was equipped with an auxiliary power unit (APU) and capable of unassisted starts. The aircraft's original APU could provide only minimal electric power and pressurized air for both aircraft cooling and for the engines' pneumatic starters. A newer, more powerful APU could provide full electrical service to the aircraft. The APU itself was started from a hydraulic accumulator by pulling a mechanical handle in the cockpit. The APU accumulator was fed from the primary hydraulic system, but could also be pumped up manually (with much effort) from the cockpit.

All crew members sit on forward-facing, upward-firing Douglas Escapac zero-zero ejection seats. In "group eject" mode, initiating ejection from either front seat ejects the entire crew in sequence, with the back seats ejecting 0.5 seconds before the front in order to provide safe separation. The rear seats are capable of self ejection, and the ejection sequence includes a pyrotechnic charge that stows the rear keyboard trays out of the occupants' way immediately before ejection. Safe ejection requires the seats to be weighted in pairs, and when flying with a single crewman in the back the unoccupied seat is fitted with ballast blocks.

At the time it entered the fleet, the S-3 introduced an unprecedented level of systems integration. Previous ASW aircraft like the Lockheed P-3 Orion and S-3's predecessor, the Grumman S-2 Tracker, featured separate instrumentation and controls for each sensor system. Sensor operators often monitored paper traces, using mechanical calipers to make precise measurements and annotating data by writing on the scrolling paper. Beginning with the S-3, all sensor systems were integrated through a single General Purpose Digital Computer (GPDC). Each crew station had its own display, and the COTAC, TACCO and SENSO displays were Multi-Purpose Displays (MPD), capable of displaying data from any of a number of systems. This new level of integration allowed the crew to consult with each other by examining the same data at multiple stations simultaneously, to manage workload by assigning responsibility for a given sensor from one station to another, and to easily combine clues from each sensor to classify faint targets. Because of this, the four-man S-3 was considered roughly equivalent in capability to the much larger P-3 with a crew of 12.

The aircraft has two underwing hard points that can be used to carry fuel tanks, general purpose and cluster bombs, missiles, rockets, and storage pods. It also has four internal bomb bay stations that can be used to carry general purpose bombs, aerial torpedoes, and special stores (B57 and B61 nuclear weapons). Fifty-nine sonoboys chutes are fitted, as well as a dedicated Search and Rescue (SAR) chute. The S-3 is fitted with the ALE-39 countermeasure system and can carry up to 90 rounds of chaff, flares, and expendable jammers (or a combination of all) in three dispensers. A retractable magnetic anomaly detector (MAD) Boom is fitted in the tail.

In the late 1990s, the S-3B's role was changed from anti-submarine warfare (ASW) to anti-surface warfare (ASuW). At that time, the MAD Boom was removed, along with several hundred pounds of submarine detection electronics. With no remaining sonoboy processing capability, most of the sonoboy chutes were faired over with a blanking plate.

On 20 February 1974, the S-3A officially became operational with the Air Antisubmarine Squadron FORTY-ONE (VS-41), the "Shamrocks," at NAS North Island, California, which served as the initial S-3 Fleet Replacement Squadron (FRS) for both the Atlantic and Pacific Fleets until a separate Atlantic Fleet FRS, VS-27, was established in the 1980s. The first operational cruise of the S-3A took place in 1975 with the VS-21 "Fighting Redtails" aboard USS John F. Kennedy.

Starting in 1987, some S-3As were upgraded to S-3B standard with the addition of a number of new sensors, avionics, and weapons systems, including the capability to launch the AGM-84 Harpoon anti-ship missile. The S-3B could also be fitted with "buddy stores" external fuel tanks that allowed the Viking to refuel other aircraft. In July 1988, VS-30 became the first fleet squadron to receive the enhanced capability Harpoon/ISAR equipped S-3B, based at NAS Cecil Field in Jacksonville, FL. 16 S-3As were converted to ES-3A Shadows for carrier-based electronic intelligence (ELINT) duties. Six aircraft, designated US-3A, were converted for a specialized utility and limited cargo COD requirement. Plans were also made to develop the KS-3A carrier-based tanker aircraft to replace the retired KA-6D Intruder, but this program was ultimately cancelled after the conversion of just one early development S-3A.

With the collapse of the Soviet Union and the breakup of the Warsaw Pact, the Soviet-Russian submarine threat was perceived as much reduced, and the Vikings had the majority of their antisubmarine warfare equipment removed. The aircraft's mission subsequently changed to sea surface search, sea and ground attack, over-the-horizon targeting, and aircraft refueling.

As a result, crews were typically limited to one Naval Aviator in the pilot seat and one Naval Flight Officer (NFO) in the copilot's seat, although the addition of an additional crewmember in the TACCO seat was not unusual for certain missions. To reflect these new missions the Viking squadrons were redesignated from "Air Antisubmarine Warfare Squadrons" to "Sea Control Squadrons." Prior to the aircraft's retirement from front-line fleet use aboard US aircraft carriers, a number of upgrade programs were implemented. These include the Carrier Airborne Inertial Navigation System II (CAINS II) upgrade, which replaced older inertial navigation hardware with ring laser gyroscopes and additional GPS systems, and added electronic flight instruments (EFI). The Maverick Plus System (MPS) added the capability to employ the AGM-65E laser-guided or AGM-65F infrared-guided AGM-65 Maverick air-to-surface missile, and the AGM-84H/K Stand-off Land Attack Missile Expanded Response (SLAM/ER). The SLAM/ER is a GPS/inertial/infrared guided cruise missile derived from the AGM-84 Harpoon that can be controlled by the aircrew in the terminal phase of flight if an AWW-13 data link pod is carried by the aircraft. The S-3B saw extensive service during the 1991 Gulf War, performing attack, tanker, and ELINT duties, and launching ADM-141 TALD decoys. The aircraft also participated in the Yugoslav wars in the 1990s and in Operation Enduring Freedom in 2001.

The first ES-3A was delivered in 1991, entering service after two years of testing. The Navy established two squadrons of eight ES-3A aircraft each in both the Atlantic and Pacific Fleets to provide detachments of typically two aircraft, ten officers, and 55 enlisted aircrew, maintenance and support personnel (which comprised/supported four complete aircrews) to deploying carrier air wings. The Pacific Fleet squadron, Fleet Air Reconnaissance Squadron FIVE (VQ-5), the "Sea Shadows," was originally based at the former NAS Agana, Guam but later relocated to NAS North Island in San Diego, California with the Pacific Fleet S-3 Viking squadrons when NAS Agana closed in 1995 as a result of a 1993 Base Realignment and Closure (BRAC) decision. The Atlantic Fleet squadron, the VQ-6 "Black Ravens," were originally based with all Atlantic Fleet S-3 Vikings at the former NAS Cecil Field in Jacksonville, Florida, but later moved to NAS Jacksonville, approximately 10 miles (16 km) to the east, when NAS Cecil Field was closed in 1999 as a result of the same 1993 BRAC decision that closed NAS Agana.

The ES-3A operated primarily with carrier battle groups, providing organic 'Indications and Warning' support to the group and joint theater commanders. In addition to their warning and reconnaissance roles, and their extraordinarily stable handling characteristics and range, Shadows were a preferred recovery tanker (aircraft that provide refueling for returning aircraft). They averaged over 100 flight hours per month while deployed. Excessive utilization caused earlier than expected equipment replacement when Naval aviation funds were limited, making them an easy target for budget-driven decision makers. In 1999, both ES-3A squadrons and all 16 aircraft were decommissioned and the ES-3A inventory placed in Aerospace Maintenance and Regeneration Group (AMARG) storage at Davis-Monthan AFB, Arizona.

Though a proposed airframe known as the Common Support Aircraft was once advanced as a successor to the S-3, E-2 and C-2, this plan failed to materialize. As the surviving S-3 airframes were forced into sundown retirement, a Lockheed Martin full scale fatigue test was performed and extended the service life of the aircraft by approximately 11,000 flight-hours. This supported Navy plans to retire all Vikings from front-line Fleet service by 2009 so new strike fighter and multi-mission aircraft could be introduced to recapitalize the aging Fleet inventory, with former Viking missions assumed by other fixed-wing and rotary-wing aircraft.

Lockheed U-2

The Lockheed U-2, nicknamed "Dragon Lady", is a single-engine, very high-altitude reconnaissance aircraft operated by the United States Air Force (USAF) and previously flown by the Central Intelligence Agency (CIA). It provides day and night, very high-altitude (70,000 feet / 21,000 m), all-weather intelligence gathering. The aircraft is also used for electronic sensor research and development, satellite calibration, and satellite data validation.

The U-2 has prominently featured in several events during the Cold War, at stages of which U-2s commonly overflew the Soviet Union, the People's Republic of China, North Vietnam, and Cuba. On 1 May 1960, CIA pilot Gary Powers was shot down while flying a U-2 over Soviet territory. During the Cuban Missile Crisis, on 27 October 1962, a U-2 piloted by Major Rudolf Anderson, Jr was shot down over Cuba by two SA-2 Guideline surface-to-air missiles.

A Lockheed TR-1 in flight

In the early 1950s, with Cold War tensions on the rise, the U.S. military desired better strategic reconnaissance to help determine Soviet capabilities and intentions. The existing reconnaissance aircraft, primarily bombers converted for reconnaissance duty, were vulnerable to anti-aircraft artillery, missiles, and fighters. It was thought an aircraft that could fly at 70,000 feet (21,000 m) would be beyond the reach of Soviet fighters, missiles, and even radar. This would allow overflights to take aerial photographs.

Under the code name "Bald Eagle", the Air Force gave contracts to Bell Aircraft, Martin Aircraft, and Fairchild Engine and Airplane to develop proposals for the new reconnaissance aircraft. Officials at Lockheed Aircraft Corporation heard about the project and asked aeronautical engineer Clarence "Kelly" Johnson to come up with a design.

Johnson was a brilliant designer, responsible for the P-38, and the P-80. He was also known for completing projects ahead of schedule, working in a separate division of the company jokingly called the Skunk Works.

Johnson's design, called the CL-282, married long glider-like wings to the fuselage of another of his designs, the Lockheed F-104 Starfighter. To save weight, his initial design did not have conventional landing gear, taking off from a dolly and landing on skids. The design was rejected by the Air Force, but caught the attention of several civilians on the review panel, notably Edwin Land, the father of instant photography. Land proposed to CIA director Allen Dulles that his agency should fund and operate this aircraft. After a meeting with President Eisenhower, Lockheed received a $22.5 million contract for the first 20 aircraft. It was renamed the U-2, with the "U" referring to the deliberately vague designation "utility". The CIA assigned the cryptonym "Aquatone" to the project, with the Air Force using the name "Oilstone" for their support to the CIA.

The first flight occurred at the Groom Lake test site (Area 51) on August 1, 1955, during what was only intended to be a high-speed taxi run. The sailplane-like wings were so efficient that the aircraft jumped into the air at 70 knots (81 mph; 130 km/h).

James Baker developed the optics for a large-format camera to be used in the U-2 while working for Perkin-Elmer. These new cameras had a resolution of 2.5 feet (76 cm) from an altitude of 60,000 feet (18,000 m). Balancing is so critical on the U-2 that the camera had to use a split film, with reels on one side feeding forward while those on the other side feed backward, thus maintaining a balanced weight distribution through the whole flight.

When the first overflights of the Soviet Union were tracked by radar, the CIA initiated Project Rainbow to reduce the U-2's radar cross section. This effort ultimately proved unsuccessful, and work began on a follow-on aircraft, which resulted in the Lockheed A-12 Oxcart.

Manufacturing was restarted in the 1980s to produce TR-1, an updated and modernized design of the U-2.

The unique design that gives the U-2 its remarkable performance also makes it a difficult aircraft to fly. It was designed and manufactured for minimum airframe weight, which results in an aircraft with little margin for error. Most aircraft were single-seat versions, with only five two-seat trainer versions known to exist. Early U-2 variants were powered by Pratt & Whitney J57 turbojet engines. The U-2C and TR-1A variants used the more powerful Pratt & Whitney J75 turbojet. The U-2S and TU-2S variants incorporated the even more powerful General Electric F118 turbofan engine.

High-aspect-ratio wings give the U-2 some glider-like characteristics, with a lift-to-drag ratio estimated in the high 20s. To maintain their operational ceiling of 70,000 feet (21,000 m), the U-2A and U-2C models (no longer in service) must fly very near their maximum speed. The aircraft's stall speed at that altitude is only 10 knots (12 mph; 19 km/h) below its maximum speed. This narrow window was referred to by the pilots as the "coffin corner". For 90% of the time on a typical mission the U-2 was flying within only five knots above stall, which might cause a decrease in altitude likely to lead to detection, and additionally might overstress the lightly built airframe.

The U-2's flight controls are designed around the normal flight envelope and altitude at which the aircraft was intended to fly. The controls provide feather-light control response at operational altitude. However, at lower altitudes, the higher air density and lack of a power-assisted control system makes the aircraft very difficult to fly.

Control inputs must be extreme to achieve the desired response in flight attitude, and a great deal of physical strength is needed to operate the controls in this manner.

The U-2 is very sensitive to crosswinds which, together with its tendency to float over the runway, makes the U-2 notoriously difficult to land. As the aircraft approaches the runway, the cushion of air provided by the high-lift wings in ground effect is so pronounced that the U-2 will not land unless the wing is fully stalled. To assist the pilot, the landing U-2 is paced by a chase car (usually a "souped-up" performance model including a Ford Mustang SSP, Chevrolet Camaro B4C, Pontiac GTO, and the Pontiac G8 GT) with an assistant (another U-2 pilot) who "talks" the pilot down by calling off the declining height of the aircraft in feet as it decreases in airspeed.

Instead of the typical tricycle landing gear, the U-2 uses a bicycle configuration with a forward set of main wheels located just behind the cockpit, and a rear set of main wheels located behind the engine. The rear wheels are coupled to the rudder to provide steering during taxiing. To maintain balance while taxiing, two auxiliary wheels, called "pogos" are added for takeoff. These fit into sockets underneath each wing at about mid-span, and fall off during takeoff. To protect the wings during landing, each wingtip has a titanium skid. After the U-2 comes to a halt, the ground crew re-installs the pogos one wing at a time, then the aircraft taxis to parking.

Because of the high operating altitude and the U-2's cockpit partial pressurization, equivalent to 29,000 feet, the pilot must wear the equivalent of a space suit. The suit delivers the pilot's oxygen supply and emergency protection in case cabin pressure is lost at altitude (the cabin provides pressure equivalent to about 29,000 feet / 8,800 meters). To prevent hypoxia and decrease the chance of decompression sickness, pilots don a full pressure suit and begin breathing 100% oxygen one hour prior to launch to remove nitrogen from the body; while moving from the building to the aircraft they breathe from a portable oxygen supply. Despite these measures, more than a dozen pilots have suffered the effects of decompression sickness since 2001; nine were reported to have suffered permanent brain damage. Inability to read fine print and inability to recognize the landing field upon return from a mission were among the symptoms. These cases were rare prior to 2001. Factors increasing the risk of illness have included longer mission durations, and more activity in the cockpit. For reconnaissance missions over the Soviet Union or China, the pilot's duties were limited to maintaining flight path while the cameras took pictures; over Afghanistan, U-2 pilots communicated in real time with ground troops and carried out more tasks in the cockpit, increasing their bodies' oxygen requirements and the risk of nitrogen bubbles forming in capillaries. The Air Force is studying the issue; U-2 pilots now exercise during oxygen pre-breathing. Among other remedies proposed is increasing cockpit pressurization to an equivalent of 15,000 feet.

The aircraft carries a variety of sensors in the nose, Q-bay (behind the cockpit, also known as the camera bay), and wing pods. The U-2 is capable of simultaneously collecting signals, imagery intelligence and air samples. Imagery intelligence sensors include either wet film photo, electro-optic or radar imagery – the latter from the Raytheon ASARS-2 system. It can use both line-of-sight and beyond-line-of-sight data links. One of the most unusual instruments in the newest version of the U-2 is the off-the-shelf Sony video camera that functions as a digital replacement for the purely optical viewsight (an upside down periscope-like viewing device) that was used in older variants to get a precise view of the terrain directly below the aircraft, especially during landing.

Lockheed Martin Supersonic Design Concept

This future aircraft design concept for supersonic flight over land comes from the team led by the Lockheed Martin Corporation.

The team's simulation shows possibility for achieving overland flight by dramatically lowering the level of sonic booms through the use of an "inverted-V" engine-under wing configuration.

Other revolutionary technologies help achieve range, payload and environmental goals.

This supersonic cruise concept is among the designs presented in April 2010 to the NASA Aeronautics Research Mission Directorate for its NASA Research Announcement-funded studies into advanced aircraft that could enter service in the 2030-2035 time-frame.

Future aircraft design concept for supersonic flight over land

Bibliography

1. Aronstein, David C. and Albert C. Piccirillo. HAVE BLUE and the F-117A. Reston, Virginia: AIAA, 1997. ISBN 1-56347-245-7.

2. Bakse, Colin. Airlift Tanker: History of U.S. Airlift and Tanker Forces. New York: Turner Publishing, 1995. ISBN 1-56311-125-X.

3. Borgu, Aldo. A Big Deal: Australia's Future Air Combat Capability. Canberra: Australian Strategic Policy Institute, 2004. ISBN 1-920722-25-4.

4. Borman, Martin W. Lockheed C-130 Hercules. Marlborough, UK: Crowood Press, 1999. ISBN 978-1-86126-205-9.

5. Chant, Christopher. Air War in the Gulf 1991. Oxford, UK: Osprey Publishing, 2001. ISBN 1-84176-295-4.

6. Crocker, H.W. III. Don't Tread on Me. New York: Crown Forum, 2006. ISBN 978-1-4000-5363-6.

7. Crosby, Francis. Fighter Aircraft. London: Lorenz Books, 2002. ISBN 0-7548-0990-0.

8. Diehl, Alan E., PhD, Former Senior USAF Safety Scientist. Silent Knights: Blowing the Whistle on Military Accidents and Their Cover-ups. Dulles, Virginia: Brassey's Inc., 2002. ISBN 1-57488-544-8.

9. Donald, David, ed. "Lockheed C-130 Hercules". The Complete Encyclopedia of World Aircraft. New York: Barnes & Noble Books, 1997. ISBN 0-7607-0592-5.

10. Dunstan, Simon. The Yom Kippur War: the Arab-Israeli War of 1973. Oxford, UK: Osprey Publishing, 2007. ISBN 1-84603-288-1.

11. Eden, Paul. "Lockheed C-130 Hercules". Encyclopedia of Modern Military Aircraft. London: Amber Books, 2004. ISBN 1-904687-84-9.

12. Frawley, Gerard. The International Directory of Military Aircraft, 2002/03. Fyshwick, ACT, Australia: Aerospace Publications Pty Ltd, 2002. ISBN 1-875671-55-2.

13. Goodall, James C. "The Lockheed F-117A Stealth Fighter". America's Stealth Fighters and Bombers: B-2, F-117, YF-22 and YF-23. St. Paul, Minnesota: Motorbooks International, 1992. ISBN 0-87938-609-6.

14. Gunston, Bill. Yakovlev Aircraft since 1924. London: Putnam Aeronautical Books, 1997. ISBN 1-55750-978-6.

15. Irving, Clive. Wide Body: The Triumph of the 747. New York: W. Morrow, 1993. ISBN 0-688-09902-5.

16. Jenkins, Dannis J. Lockheed Secret Projects: Inside the Skunk Works. St. Paul, Minnesota: Zenith Imprint, 2001. ISBN 0-7603-0914-0.

17. Johnsen, Frederick A.. Lockheed C-141 Starlifter. North Branch, Minnesota: Specialty Press, 2005. ISBN 1-58007-080-9.

18. Keijsper, Gerald. Lockheed F-35 Joint Strike Fighter. London: Pen & Sword Aviation, 2007. ISBN 978-1-84415-631-3.

19. Lippincott, Richard. C-5 Galaxy in Action. Carrollton, Texas: Squadron/Signal Publications, 2006. ISBN 0-89747-504-6.

20. Logan, Don. Lockheed F-117 Nighthawks: A Stealth Fighter Roll Call. Atglen, Pennsylvania: Schiffer Publishing, 2009. ISBN 978-0-7643-3242-5.

21. Miller, David. Conflict Iraq: Weapons and Tactics of the US and Iraqi Forces. St. Paul, Minnesota: Zenith Imprint, 2003. ISBN 0-7603-1592-2.

22. Nalty, Bernard C. Winged Shield, Winged Sword 1950-1997: A History of the United States Air Force. Minerva Group, 2003. ISBN 1-4102-0902-4.

23. Norton, Bill. Lockheed Martin C-5 Galaxy. North Branch, Minnesota: Specialty Press, 2003. ISBN 1-58007-061-2.

24. Ogden, Bob. Aviation Museums and Collections of North America (2 ed.). Tonbridge, Kent: Air Britain (Historians) Ltd, 2011. ISBN 978-0-85130-427-4.

25. Pace, Steve. X-Fighters: USAF Experimental and Prototype Fighters, XP-59 to YF-23. St. Paul, Minnesota: Motorbooks International, 1991. ISBN 0-87938-540-5.

26. Philips, Warren F. Mechanics of Flight. Hoboken, New Jersey: John Wiley and Sons, 2004. ISBN 0-471-33458-8.

27. Polmar, Norman. The Naval Institute Guide to the Ships and Aircraft of the U.S. Fleet. Annapolis, Maryland: Naval Institute Press, 2005. ISBN 978-1-59114-685-8.

28. Reed, Chris. Lockheed C-130 Hercules and Its Variants. Atglen, Pennsylvania: Schiffer Publishing, 1999. ISBN 978-0-7643-0722-5.

29. Reed, Chris. Lockheed C-5 Galaxy. Atglen, Pennsylvania: Schiffer Publishing, 2000. ISBN 0-7643-1205-7.

30. Richardson, Doug. Stealth Warplanes. New York: Salamander Books Ltd, 2001. ISBN 0-7603-1051-3.

31. Spick, Mike. The Illustrated Directory of Fighters. London: Salamander, 2002. ISBN 1-84065-384-1.

32. Tillman, Barrett. What We Need: Extravagance and Shortages in America's Military. Zenith Imprint, 2007. ISBN -076032-869-2.

33. Veronico, Nick and Jim Dunn. 21st Century U.S. Air Power. St. Paul, Minnesota: Zenith Imprint, 2004. ISBN 0-7603-2014-4.

34. Williams, Mel, ed. "Lockheed Martin F-22A Raptor". Superfighters: The Next Generation of Combat Aircraft. London: AIRtime Publishing Inc., 2002. ISBN 1-880588-53-6.

35. Winchester, Jim. "Lockheed Martin X-35/F-35 JSF". Concept Aircraft: Prototypes, X-Planes and Experimental Aircraft. Kent, UK: Grange Books plc., 2005. ISBN 1-59223-480-1.

www.ingramcontent.com/pod-product-compliance
Lightning Source LLC
Chambersburg PA
CBHW051019180526
45172CB00002B/407